Understanding Physical Optics

Understanding Physical Optics

Frank Bailey

Willford Press
www.willfordpress.com

Published by Willford Press,
118-35 Queens Blvd., Suite 400,
Forest Hills, NY 11375, USA

ISBN: 978-1-68285-961-2

Cataloging-in-Publication Data

Understanding physical optics / Frank Bailey.
p. cm.
Includes bibliographical references and index.
ISBN 978-1-68285-961-2
1. Physical optics. 2. Optics. I. Bailey, Frank.
QC355.3 .U53 2020
535--dc23

For information on all Willford Press publications
visit our website at www.willfordpress.com

Willford Press

Printed in China.

Contents

Preface

Optics is the branch of physics which deals with the properties and behavior of light. It is divided into two subfields, physical optics and geometrical optics. The branch of optics which focuses on the phenomenon of interference, diffraction and polarization is known as physical optics. It is an intermediate method between full wave electromagnetism and geometric optics. According to this discipline, light is believed to propagate in the form of a wave, rather than travel in a straight line. This is the reason due to which it is also known as wave optics. The objective of this book is to give a general view of the different areas of physical optics. The extensive content of this book provides the readers with a thorough understanding of the subject. Students, researchers, experts and all associated with this discipline will benefit alike from this book.

A detailed account of the significant topics covered in this book is provided below:

Chapter 1- The branch of physics that deals with the study of properties and behavior of light and its interaction with matter is termed as optics. It studies the behavior of visible, ultraviolet and visible light and is mainly categorized into two types – ray optics and wave optics. This is an introductory chapter which will briefly introduce all the aspects of optics.

Chapter 2- Interference of light is defined as the phenomenon of reinforcing of two light waves in order to form a greater wave. Fresnel biprism, Lloyd's mirror arrangement, Newton's rings, multiple-beam interferometry, coherence, Fourier transform spectroscopy, etc. are a few of the concepts that fall under its domain. All these concepts of interference of light have been carefully analyzed in the chapter.

Chapter 3- Diffraction is defined as the bending of light wave around the edge of an object which depends on the wavelength of light wave. It includes Huygens' principle, Fraunhofer diffraction, single and double slit diffraction, diffraction grating, Fresnel diffraction, etc. The topics elaborated in this chapter will help in gaining a better perspective about these aspects of refraction of light.

Chapter 4- Reflection is defined as the bouncing back of light from a surface. Refraction is defined as the bending of light as it passes from one medium to another. This chapter has been carefully written to provide an easy understanding of the concepts of reflection and refraction such as total internal reflection, image formation by lenses, etc. with respect to physical optics.

Chapter 5- Polarization is the property of transverse wave that defines the geometrical orientation of the oscillations. Double refraction deals with the anisotropic materials that depend on the polarization and direction of propagation of light. This chapter closely examines all the aspects of polarization and double refraction to provide an extensive understanding of the subject.

I would like to make a special mention of my publisher who considered me worthy of this opportunity and also supported me throughout the process. I would also like to thank the editing team at the back-end who extended their help whenever required.

Frank Bailey

1

Optics: An Introduction

The branch of physics that deals with the study of properties and behavior of light and its interaction with matter is termed as optics. It studies the behavior of visible, ultraviolet and visible light and is mainly categorized into two types – ray optics and wave optics. This is an introductory chapter which will briefly introduce all the aspects of optics.

Light

Light is an electromagnetic radiation that can be detected by the human eye. Electromagnetic radiation occurs over an extremely wide range of wavelengths, from gamma rays with wavelengths less than about 1×10^{-11} metre to radio waves measured in metres. Within that broad spectrum the wavelengths visible to humans occupy a very narrow band, from about 700 nanometres (nm; billionths of a metre) for red light down to about 400 nm for violet light. The spectral regions adjacent to the visible band are often referred to as light also, infrared at the one end and ultraviolet at the other. The speed of light in a vacuum is a fundamental physical constant, the currently accepted value of which is exactly 299,792,458 metres per second, or about 186,282 miles per second.

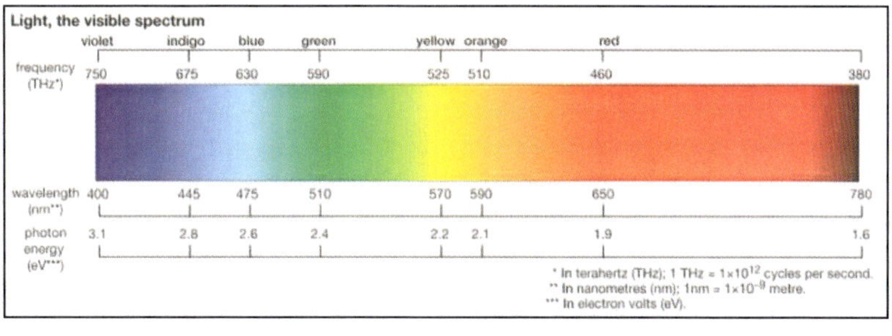

When white light is spread apart by a prism or a diffraction grating, the colours of the visible spectrum appear. The colours vary according to their wavelengths. Violet has the highest frequencies and shortest wavelengths, and red has the lowest frequencies and the longest wavelengths.

No single answer to the question "What is light?" satisfies the many contexts in which light is experienced, explored, and exploited. The physicist is interested in the physical properties of light, the artist in an aesthetic appreciation of the visual world. Through the sense of sight, light is a primary tool for perceiving the world and communicating within it. Light from the Sun warms the Earth, drives global weather patterns, and initiates the life-sustaining process of photosynthesis. On the grandest scale, light's interactions with matter have helped shape the structure of the universe. Indeed, light provides a window on the universe, from cosmological to atomic scales. Almost all of the information about the rest of the universe reaches Earth in the form of electromagnetic radiation. By interpreting that radiation, astronomers can glimpse the earliest epochs of the universe, measure the general expansion of the universe, and determine the chemical composition of stars and the interstellar medium. Just as the invention of the telescope dramatically broadened exploration of the universe, so too the invention of the microscope opened the intricate world of the cell. The analysis of the frequencies of light emitted and absorbed by atoms was a principal impetus for the development of quantum mechanics. Atomic and molecular spectroscopies continue to be primary tools for probing the structure of matter, providing ultrasensitive tests of atomic and molecular models and contributing to studies of fundamental photochemical reactions.

The Sun shining from behind clouds.

Light transmits spatial and temporal information. This property forms the basis of the fields of optics and optical communications and a myriad of related technologies, both mature and emerging. Technological applications based on the manipulations of light include lasers, holography, and fibre-optic telecommunications systems.

In most everyday circumstances, the properties of light can be derived from the theory of classical electromagnetism, in which light is described as coupled electric and magnetic fields propagating through space as a traveling wave. However, this wave theory, developed in the mid-19th century, is not sufficient to explain the properties of light at very low intensities. At that level a quantum theory is needed to explain the characteristics of light and to explain the interactions of light with atoms and molecules. In its simplest form, quantum theory describes light as consisting of discrete packets of energy, called photons. However, neither a classical wave model nor a classical particle model correctly describes light; light has a dual nature that is revealed only in quantum mechanics. This surprising wave-particle duality is shared by all of the primary constituents of nature (e.g., electrons have both particle-like and wavelike aspects). Since the mid-20th century, a more comprehensive theory of light, known as quantum electrodynamics (QED), has been regarded by physicists as complete. QED combines the ideas of classical electromagnetism, quantum mechanics, and the special theory of relativity.

Optics

Optics is the science concerned with the genesis and propagation of light, the changes that it undergoes and produces, and other phenomena closely associated with it. There are two major branches of optics, physical and geometrical. Physical optics deals primarily with the nature and properties of light itself. Geometrical optics has to do with the principles that govern the image-forming properties of lenses, mirrors, and other devices that make use of light. It also includes optical data processing, which involves the manipulation of the information content of an image formed by coherent optical systems.

Originally, the term optics was used only in relation to the eye and vision. Later, as lenses and other devices for aiding vision began to be developed, these were naturally called optical instruments, and the meaning of the term optics eventually became broadened to cover any application of light, even though the ultimate receiver is not the eye but a physical detector, such as a photographic plate or a television camera. In the 20th century optical methods came to be applied extensively to regions of the electromagnetic radiation spectrum not visible to the eye, such as X-rays, ultraviolet, infrared, and microwave radio waves, and to this extent these regions are now often included in the general field of optics.

Geometrical Optics

The Optical Image

An optical image may be regarded as the apparent reproduction of an object by a lens or mirror system, employing light as a carrier. An entire image is generally produced simultaneously, as by the lens in a camera, but images may also be generated sequentially by point-by-point scanning, as in a television system or in the radio transmission of pictures across long distances in space. Nevertheless, the final detector of all images is invariably the human eye, and, whatever means is used to transmit and control the light, the final image must either be produced simultaneously or scanned so rapidly that the observer's persistence of vision will give him the mental impression of a complete image covering a finite field of view. For this to be effective the image must be repeated (as in motion pictures) or scanned (as in television) at least 40 times a second to eliminate flicker or any appearance of intermittency.

Light Rays, Waves and Wavelets

A single point of light, which may be a point in an extended object, emits light in the form of a continually expanding train of waves, spherical in shape and centred about the point of light. It is, however, often much more convenient to regard an object point as emitting fans of rays, the rays being straight lines everywhere perpendicular to the waves. When the light beam is refracted by a lens or reflected by a mirror, the curvature of the waves is changed, and the angular divergence of the ray bundle is similarly changed in such a way that the rays remain everywhere perpendicular to the waves. When aberrations are present, a convergent ray bundle does not shrink to a perfect point, and the emerging waves are then not truly spherical.

In 1690 Christiaan Huygens, a Dutch scientist, postulated that a light wave progresses because each point in it becomes the centre of a little wavelet travelling outward in all directions at the speed of light, each new wave being merely the envelope of all these expanding wavelets. When the wavelets reach the region outside the outermost rays of the light beam, they destroy each other by mutual interference wherever a crest of one wavelet falls upon a trough of another wavelet. Hence, in effect, no waves or wavelets are allowed to exist outside the geometrical light beam defined by the rays. The normal destruction of one wavelet by another, which serves to restrict the light energy to the region of the rectilinear ray paths, however, breaks down when the light beam strikes an opaque edge, for the edge then cuts off some of the interfering wavelets, allowing others to exist, which diverge slightly into the shadow area. This phenomenon is called diffraction, and it gives rise to a complicated fine structure at the edges of shadows and in optical images.

The Inhole Camera

An excellent example of the working of the wavelet theory is found in the well-known pinhole camera. If the pinhole is large, the diverging geometrical pencil of rays leads to a blurred image, because each point in the object will be projected as a finite circular patch of light on the film. The spreading of the light at the boundary of a large pinhole by diffraction is slight. If the pinhole is made extremely small, however, the geometrical patch then becomes small, but the diffraction spreading is now great, leading once more to a blurred picture. There are thus two opposing effects present, and at the optimum hole size the two effects are just equal. This occurs when the hole diameter is equal to the square root of twice the wavelength (λ) times the distance (f) between the pinhole and film—i.e., Square root of $\sqrt{2\lambda}$ f. For f = 100 millimetres and λ = 0.0005 millimetre, the optimum hole size becomes 0.32 millimetre. This is not very exact, and a 0.4-millimetre hole would probably be just as good in practice. A pinhole, like a camera lens, can be regarded as having an f-number, which is the ratio of focal length to aperture. In this example, the f-number is 100/0.32 = 310, designated f/310. Modern camera lenses have much greater apertures, in order to achieve light-gathering power, of around f/1.2–f/5.6.

Resolution and the Airy disk

When a well-corrected lens is used in place of a pinhole, the geometrical ray divergence is eliminated by the focussing action of the lens, and a much larger aperture may be employed; in that case the diffraction spreading becomes small indeed. The image of a point formed by a perfect lens is a minute pattern of concentric and progressively fainter rings of light surrounding a central dot, the whole structure being called the Airy disk after George Biddell Airy, an English astronomer, who first explained the phenomenon in 1834. The Airy disk of a practical lens is small, its diameter being approximately equal to the *f*-number of the lens expressed in microns (0.001 millimetre). The Airy disk of an *f*/4.5 lens is therefore about 0.0045 millimetre in diameter (ten times the wavelength of blue light). Nevertheless, the Airy disk formed by a telescope or microscope objective can be readily seen with a bright point source of light if a sufficiently high eyepiece magnification is used.

The finite size of the Airy disk sets an inevitable limit to the possible resolving power of a visual instrument. Rayleigh found that two adjacent and equally bright stars can just be resolved if the image of one star falls somewhere near the innermost dark ring in the Airy disk of the other star; the resolving power of a lens can therefore be regarded as about half the *f*-number of the lens expressed in microns. The angular resolution of a telescope is equal to the angle subtended by the least resolvable image separation at the focal length of the objective, the light-gathering lens. This works out at about four and a half seconds of arc divided by the diameter of the objective in inches.

The Rayleigh Limit

When a perfect lens forms an image of a point source of light, the emerging wave is a sphere centred about the image point. The optical paths from all points on the wave to the image are therefore equal, so that the expanding wavelets are all in phase (vibrating in unison) when they reach the image. In an imperfect lens, however, because of the presence of aberrations, the emerging wave is not a perfect sphere, and the optical paths from the wave to the image point are then not all equal. In such a case some wavelets will reach the image as a peak, some as a trough, and there will be much destructive interference leading to the formation of a sizable patch of light, much different from the minute Airy disk characteristic of a perfectly corrected lens. In 1879 Rayleigh studied the effects of phase inequalities in a star image and came to the conclusion that an image will not be seriously degraded unless the path differences between one part of the wave and another exceed one-quarter of the wavelength of light. As this difference represents only 0.125 micron (5×10^{-6} inch), it is evident that an optical system must be designed and constructed with almost superhuman care if it is to give the best possible definition.

Reflection and Refraction

Reflection

The use of polished mirrors for reflecting light has been known for thousands of years, and concave mirrors have long been used to form real images of distant objects. Indeed, Isaac Newton greatly preferred the use of a mirror as a telescope objective to the poor-quality lenses available in his time. Because there is no limit to the possible size of a mirror, all large telescopes today are of this type.

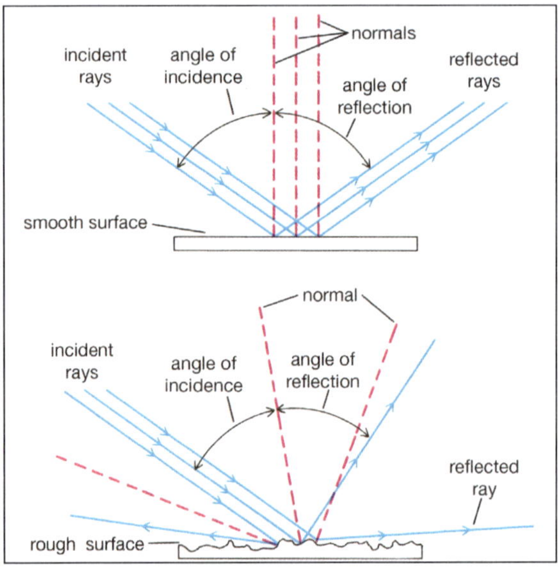

Reflection of light: In the reflection of light, the angle of incidence is equal to the angle of reflection, measured from the normal (the line perpendicular to the point of impact).

When a ray of light is reflected at a polished surface, the angle of reflection between ray and normal (the line at right angles to the surface) is exactly equal to the angle of incidence. It can be seen that a convex mirror forms a virtual image of a distant object, whereas a concave mirror forms a real image. A plane mirror forms a virtual image of near objects, as in the familiar looking glass. Plane mirrors are often used in instruments to bend a beam of light into a different direction.

The Law of Refraction

When a ray of light meets the surface of separation between two transparent media, it is sharply bent or refracted. Because rays are really only directions and have no physical existence, the passage of light waves through a surface must be considered if refraction is to be understood. Refraction effects are based on the fact that light travels more slowly in a denser medium. The ratio of the velocity of light in air to its velocity in the medium is called the refractive index of the medium for light of a particular colour or wavelength. The refractive index is higher for blue light than for light at the red end of the spectrum.

In figure, AA′ represents a plane wave of light at the instant that A′ meets the plane refracting surface A′B separating two media having refractive indices n and n′, respectively. During the time taken by the light to travel from A to B in material n, light travels from A′ to B′ in material of refractive index n′, forming the new wave BB′ in the second material, proceeding in direction BC. Hence, the relationship n′/n = AB/A′B ′ can be obtained; and dividing numerator and denominator by BA′ gives:

$$\frac{n'}{n} = \frac{AB/BA'}{A'B'/BA'} = \frac{\sin I}{\sin I'}.$$

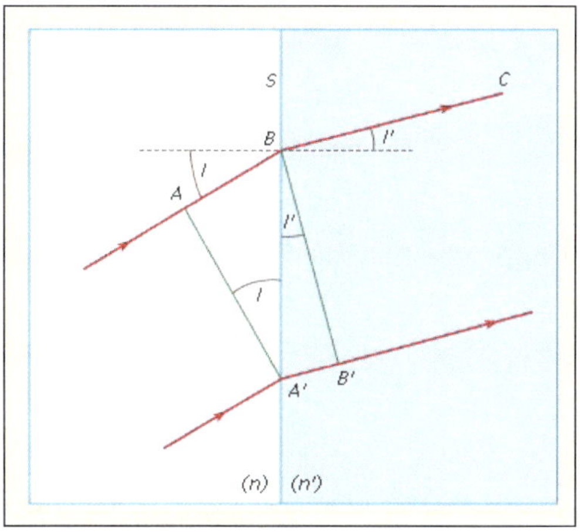

The law of refraction. Plane light wave at position AA′ in medium of index n and BB′ in medium of index n′.

The angles I and I′ are called the angle of incidence and angle of refraction between the refracting surface and the incident and refracted waves, respectively.

Returning now to the convention of considering the movement of light in terms of rays because entering and emerging rays are always perpendicular to the light waves they represent, angles I and I′ also denote the angles between the entering and emerging rays and the normal (perpendicular) line to the refracting surface at B.

Equation $\frac{n'}{n} = \frac{AB / BA'}{A'B'/ BA'} = \frac{\sin I}{\sin I'}$., known as the law of refraction, is generally written: n′ sin I' = n sin I.

Dispersion

The difference between the refractive indices of a transparent material for a specific blue light and a specific red light is known as the dispersion of the material. The usual choices of blue and red lights are the so-called "F" and "C" lines of hydrogen in the solar spectrum, named by Fraunhofer, with wavelengths 4861 and 6563 angstroms (the angstrom unit, abbreviated Å, is 10^{-8} centimetre), respectively. It is generally more significant, however, to compare the dispersion with the mean refractive index of the material for some intermediate colour such as the sodium "D" Fraunhofer line of wavelength 5893 angstroms. The dispersive power (w) of the material is then defined as the ratio of the difference between the "F" and "C" indices and the "D" index reduced by 1, or,

$$w = \frac{n_F - n_C}{n_D - 1}.$$

Hundreds of different types of optical glass are currently available from manufacturers. These may be represented graphically on a plot of mean refractive index against dispersive power.

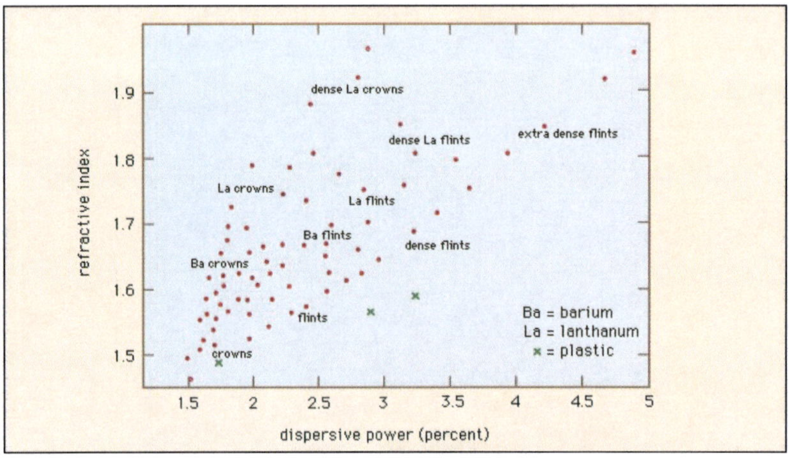

Relationships between refractive indices and dispersive powers of several representative optical glasses and plastics.

At first lenses were made from selected pieces of window glass or the glass used to make blown tableware. In the early 1800s, the manufacture of clear glass that was intended specifically for lenses began in Europe. The glass was slowly stirred in the molten state to remove striations and irregularities, and then the whole mass was cooled and broken up into suitable pieces for lens making. Subsequently, the pieces were placed in molds of the approximate size of the lens, slowly remelted to shape, and carefully annealed; i.e., allowed to cool slowly under controlled conditions to reduce strains and imperfections. Various chemicals were added in the molten state to vary the properties of the glass: addition of lead oxide, for example, was found to raise both the refractive index and the dispersive power. In 1884 it was discovered that barium oxide had the effect of raising the refractive index without increasing the dispersion, a property that proved to be of the greatest value in the design of photographic lenses known as anastigmats (lenses devoid of astigmatic aberration). In 1938 a further major improvement was achieved by the use of various rare-earth elements, and since 1950 lanthanum glass has been commonly used in high-quality photographic lenses.

The cost of optical glass varies considerably, depending on the type of glass, the precision with which the optical properties are maintained, the freedom from internal striae and strain, the number of bubbles, and the colour of the glass. Many common types of optical glass are now available in quite large pieces, but as the specifications of the glass become more stringent the cost rises and the range of available sizes becomes limited. In a small lens such as a microscope objective or a telescope eyepiece, the cost of the glass is insignificant, but in large lenses in which every millimetre of thickness may represent an additional pound in weight, the cost of the glass can be very high indeed.

Lenses can be molded successfully of various types of plastic material, polymethyl methacrylate being the most usual. Even multi-element plastic lenses have been manufactured for low-cost cameras, the negative (concave) elements being made of a high-dispersion plastic such as styrene.

Total Internal Reflection

When a ray of light emerges obliquely from glass into air, the angle of refraction between ray and normal is greater than the angle of incidence inside the glass, and at a sufficiently high obliquity the angle of refraction can actually reach 90°. In this case the emerging ray travels along the glass surface, and the sine of the angle of incidence inside the glass, known as the critical angle, is then equal to the reciprocal of the refractive index of the material. At angles of incidence greater than the critical angle, the ray never emerges, and total internal reflection occurs, for there is no measurable loss if the glass surface is perfectly clean. Dirt or dust on the surface can cause a small loss of energy by scattering some light into the air.

Light is totally internally reflected in many types of reflecting prism and in fibre optics, in which long fibres of high-index glass clad with a thin layer of lower index glass are assembled side-by-side in precise order. The light admitted into one end of each fibre is transmitted along it without loss by thousands of successive internal reflections at the interlayer between the glass and the cladding. Hence, an image projected upon one end of the bundle will be dissected and transmitted to the other end, where it can be examined through a magnifier or photographed. Many modern medical instruments, such as cystoscopes and bronchoscopes, depend for their action on this principle. Single thick fibres (actually glass rods) are sometimes used to transmit light around corners to an otherwise inaccessible location.

The hair-thin fibres used in fibre optics.

Ray-tracing Methods

Graphical Ray Tracing

In 1621 Willebrord Snell, a professor of mathematics at Leiden, discovered a simple graphical procedure for determining the direction of the refracted ray at a surface when the incident ray is given. The mathematical form of the law of refraction, equation

$\frac{n'}{n} = \frac{AB / BA'}{A'B'/ BA'} = \frac{\sin I}{\sin I'}$. Above, was announced by the French mathematician René Descartes some 16 years later.

Snell's construction is as follows: The line AP in figure represents a ray incident upon a refracting surface at P, the normal at P being PN. If the incident and refracted rays are extended to intersect any line SS parallel to the normal, the lengths PQ and PR along the rays will be proportional to the refractive indices n and n′. Hence, if PQ and the indices are known, PR can be found and the refracted ray drawn in.

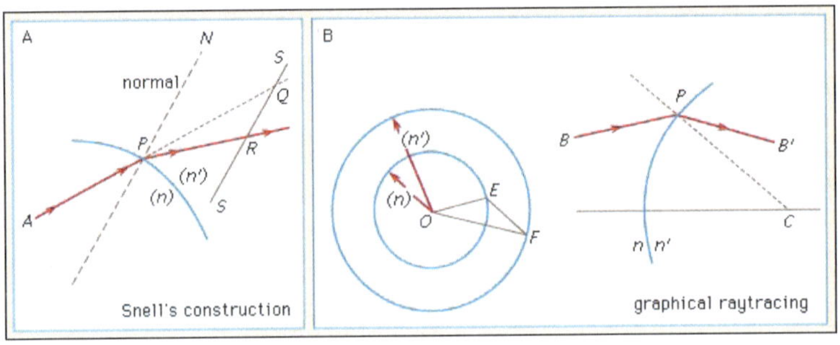

Graphic refraction procedures.

A convenient modification of Snell's construction can readily be used to trace the path of a ray through a complete lens. In figure, the incident ray BP strikes a refracting surface at *P*. The normal to the surface is *PC*. At any convenient place on the page two concentric circles are drawn about a point *O* with radii proportional to the refractive indices *n* and n′, respectively. A line *OE* is now drawn parallel to the incident ray *BP* extending as far as the circle representing the refractive index n of the medium containing the incident ray. From *E* a line is drawn parallel to the normal *PC* extending to *F* on the circle representing the refractive index *n′*. The line OF then represents the direction of the desired refracted ray, which may be drawn in at *PB′*. This process is repeated successively for all the surfaces in a lens. If a mirror is involved, the reflected ray may be found by drawing the normal line EF across the circle diagram to the incident-index circle on the other side.

Trigonometrical Ray Tracing

No graphical construction can possibly be adequate to determine the aberration residual of a corrected lens, and for this an accurate trigonometrical computation must be made and carried out to six or seven decimal places, the angles being determined to single seconds of arc or less. There are many procedures for calculating the path of a ray through a system of spherical refracting or reflecting surfaces, the following being typical: The diagram in figure represents a ray lying in the meridian plane, defined as the plane containing the lens axis and the object point. A ray in this plane is defined by its slope angle, U, and by the length of the perpendicular, Q, drawn from the vertex (A) of the surface on to the ray. By drawing a line parallel to the incident ray through the centre of curvature

C, to divide Q into two parts at N, the relation is stated as AN = r sin U, and NM = r sin I. Hence,

$$Q = r\left(\sin U + \sin I\right).$$

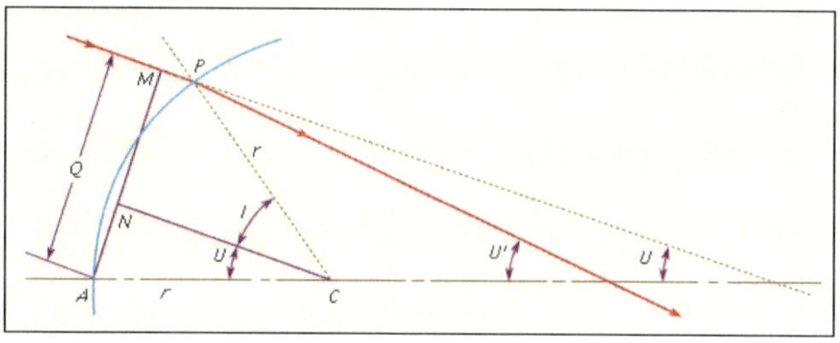

Trigonometrical ray tracing.

From this the first ray-tracing equation can be derived:

$$\sin I = \frac{Q}{r} - \sin U.$$

Applying the law of refraction, equation $Q = r\left(\sin U + \sin I\right)$, gives the second equation:

$$\sin I' = \frac{n}{n'} - \sin I.$$

Because the angle $PCA = U + I = U' + I'$, the slope of the refracted ray can be written as:

$$U' = U + l + l';$$

and, lastly, by adding primes to equation $Q = r\left(\sin U + \sin I\right)$:

$$Q' = r\left(\sin U' + \sin l'\right).$$

Having found the Q′ of the refracted ray, transfer to the next surface can be performed by:

$$Q_2 = Q_1' - d \sin U_1',$$

in which d is the axial distance from the first to the second refracting surface. After performing this calculation for all the surfaces in succession, the longitudinal distance from the last surface to the intersection point of the emergent ray with the lens axis is found by:

$$L' = \frac{Q'}{\sin U'}.$$

Corresponding but much more complicated formulas are available for tracing a skew ray, that is, a ray that does not lie in the meridian plane but travels at an angle to it.

After refraction at a surface, a skew ray intersects the meridian plane again at what is called the diapoint. By tracing the paths of a great many (100 or more) meridional and skew rays through a lens, with the help of an electronic computer, and plotting the assemblage of points at which all these rays pierce the focal plane after emerging from the lens, a close approximation to the appearance of a star image can be constructed, and a good idea of the expected performance of a lens can be obtained.

Paraxial, First-order and Imagery

In a lens that has spherical aberration, the various rays from an axial object point will in general intersect the lens axis at different points after emerging into the image space. By tracing several rays entering the lens at different heights (i.e., distances from the axis) and extrapolating from a graph connecting ray height with image position, it would be possible to infer where a ray running very close to the axis (a paraxial ray) would intersect the axis, although such a ray could not be traced directly by the ordinary trigonometrical formulas because the angles would be too small for the sine table to be of any use. Because the sine of a small angle is equal to the radian measure of the angle itself, however, a paraxial ray can be traced by reducing the ray-tracing formulas to their limiting case for small angles and thus determining the paraxial intersection point directly. When this is done, writing paraxial-ray data with lowercase letters, it is found that the Q and Q′ above both become equal to the height of incidence y, and the formulas $\sin I = \frac{Q}{r} - \sin U$ and $\sin I' = \frac{n}{n'} - \sin I$ become, in the paraxial limit:

$$i = \frac{y}{r'} - u$$

$$i' = \frac{n}{n'} i$$

$$u' = u + i - i'$$

The longitudinal distance from the last surface to the intersection point of the emerging paraxial ray with the lens axis becomes $l' = y/u'$.

Because all paraxial rays from a given object point unite at the same image point, the resulting longitudinal distance (l′) is independent of the particular paraxial ray that is traced. Any nominal value for the height of incidence, y, may therefore be adopted, remembering that it is really an infinitesimal and y is only its relative magnitude. Thus, it is clear that the paraxial angles in equations below are really only auxiliaries, and they can be readily eliminated, giving the object–image distances for paraxial rays:

$$i = \frac{y}{r'} - u$$

$$i' = \frac{n}{n'} i$$

$$u' = u + i - i'$$

$$n'(l'-r)u' = n(l-r)u$$

and

$$\frac{n'}{l'} = \frac{n}{l} + \frac{n'-n}{r}.$$

Magnification: The Optical Invariant

It is frequently as important to determine the size of an image as it is to determine its location. To obtain an expression for the magnification—that is, the ratio of the size of an image to the size of the object—the following process may be used: If an object point B lies to one side of the lens axis at a transverse distance h from it, and the image point B′ is at a transverse distance h′, then B, B′, and the centre of curvature of the surface, C, lie on a straight line called the auxiliary axis. Then, by simple proportion,

$$m = \frac{h'}{h} = \frac{l'-r}{l-r} = \frac{nu}{n'u'}$$

$$h'n'u' = hnu,$$

And the product (hnu) is invariant for all the spaces between the lens surfaces, including the object and image spaces, for any lens system of any degree of complexity. This theorem has been named after the French scientist Joseph-Louis Lagrange, although it is sometimes called the Smith-Helmholtz theorem, after Robert Smith, an English scientist, and Hermann Helmholtz, a German scientist; the product (hnu) is often known as the optical invariant. As it is easy to determine the quantities h, n, and u for the original object, it is only necessary to calculate u′ by tracing a paraxial ray in order to find the image height h′ for any lens. If the lens is used in air, as most lenses are, the refractive indices are both unity, and the magnification becomes merely m = u/u′.

The Gauss Theory of Lenses

In 1841 Gauss published a now famous treatise on optics in which he demonstrated that, so far as paraxial rays are concerned, a lens of any degree of complexity can be replaced by two principal, or nodal, points and two focal points, the distances from the principal points to their respective focal points being the focal lengths of the lens, and, furthermore, that the two focal lengths are equal to one another when the refractive indices of object and image spaces are equal, as when a lens is used in air.

The principal and focal points may be defined as follows: Figure shows a lens system of any construction, with a bundle of rays entering from the left in a direction parallel to the lens axis. After refraction by the lens each ray will cross the axis at some point, and the entering and emerging portions of each ray are then extended until they intersect at a point such as Q. The locus of all the points Q is a surface of revolution about the lens axis known as the equivalent refracting locus of the lens. The point where this locus crosses the axis is called the principal point, P_2, and the central portion of the locus in the neighbourhood of the axis, which is virtually a plane perpendicular to the axis, is called the principal plane. The point where the emerging paraxial ray crosses the axis is called the focal point F_2, the distance from P_2 to F_2 being the (posterior) focal length f'. A similar situation exists for a parallel beam of light entering from the right, giving the anterior principal point P_1, the anterior focal point F_1, and the front focal length f. For a lens in air it can be shown that the two focal lengths are equal in magnitude but opposite in direction—i.e., if F_2 is to the right of P_2, then F_1 must lie to the left of P_1, as in the case of an ordinary positive lens (one that gives a real image). In a negative lens (one that gives a virtual image), F_2 lies to the left of P_2, and the posterior focal length f' is negative.

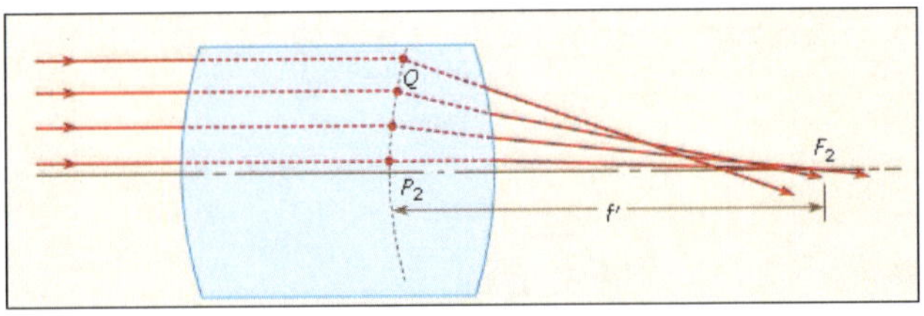

Figure: The Gauss Theory.

The relation between the distances of object and image from a lens can be easily stated if the positions of the two principal points and the two focal points are known. (In using these expressions, distances are considered positive or negative depending on whether they are measured to the right or to the left from their respective origins.) For a lens in air: (a) If the conjugate distances measured from the respective focal points are x and x', and if m is the image magnification (height of image divided by height of object), then $m = -x'/f' = f'/x$ and $xx' = -f'^2$. (b) If the conjugate distances measured from the respective principal points are p and p' and if m is the image magnification, then $m = p'/p$ and $1/p' = 1/p + 1/f'$. The Lagrange equation $h'n'u' = hnu$, requires modification for a distant object because in that case the object height h is infinite, and the slope angle u is zero. If the off-axis distance h is divided by the object distance L, and u is multiplied by L, equation $h'n'u' = hnu$, becomes $h' = (n/n')f'\phi$, in which ϕ is the angle in radians subtended by the distant object at the lens. This formula provides a means for defining focal length and for measuring the focal length of an unknown lens.

The Thin Lens

In a thin lens such as a spectacle, the two principal planes coincide within the lens, and then the conjugate distances p and p' in the formula above become the distances of object and image from the lens itself.

The focal length of a thin lens can be computed by applying the surface-conjugate formula to the two surfaces in succession, writing the l of the first surface as infinity and the l of the second surface equal to the l′ of the first surface. When this is done, the lens power *(P)* becomes:

$$\frac{n'}{l'} = \frac{n}{l} + \frac{n'-n}{r}.$$

$$P = \frac{1}{f'} = (n-1)\left(\frac{1}{r_1} - \frac{1}{r_2}\right).$$

Chromatic Aberration

Because the refractive index of glass varies with wavelength, every property of a lens that depends on its refractive index also varies with wavelength, including the focal length, the image distance, and the image magnification. The change of image distance with wavelength is known as chromatic aberration, and the variation of magnification with wavelength is known as chromatic difference of magnification, or lateral colour. Chromatic aberration can be eliminated by combining a strong lens of low-dispersion glass (crown) with a weaker lens made of high-dispersion (flint) glass. Such a combination is said to be achromatic. This method of removing chromatic aberration was discovered in 1729 by Chester Hall, an English inventor, and it was exploited vigorously in the late 18th century in numerous small telescopes. Chromatic variation of magnification can be eliminated by achromatizing all the components of a system or by making the system symmetrical about a central diaphragm. Both chromatic aberration and lateral colour are corrected in every high-grade optical system.

Longitudinal Magnification

If an object is moved through a short distance δp along the axis, then the corresponding image shift δp′ is related to the object movement by the longitudinal magnification (m). Succinctly,

$$\overline{m} = \delta p' / \delta p = m^2$$

in which m is the lateral magnification. The fact that the longitudinal magnification is equal to the square of the transverse magnification means that m is always positive; hence, if the object is moved from left to right, the image must also move from left to right. Also, if m is large, then m is very large, which explains why the depth of field (δp) of a microscope is extremely small. On the other hand, if m is small, less than one as in a camera, then m is very small, and all objects within a considerable range of distances (δp) appear substantially in focus.

Image of a Tilted Plane

If a lens is used to form an image of a plane object that is tilted relative to the lens axis, then the image will also be tilted in such a way that the plane of the object, the plane of the image, and the median plane of the lens all meet. This construction can be derived by the use of the lateral and longitudinal magnification relations just established above. With a tilted object the magnification at any point is given by the ratio of the distances of image and object from the lens at that point in the image, and, consequently, m varies progressively from one end of the image to the other. This arrangement is frequently used in view cameras equipped with "swings" to increase depth of field and in enlargers to rectify the convergence of parallel lines caused by tilting the camera, for example, in photographing tall buildings. The rule finds extensive application in photogrammetry and in the making of maps from aerial photographs.

Optical Systems

System Components

An optical system consists of a succession of elements, which may include lenses, mirrors, light sources, detectors, projection screens, reflecting prisms, dispersing devices, filters and thin films, and fibre-optics bundles.

Lenses

All optical systems have an aperture stop somewhere in the system to limit the diameter of the beams of light passing through the system from an object point. By analogy with the human eye, this limiting aperture stop is called the iris of the system, its images in the object and image spaces being called the entrance pupil and exit pupil, respectively. In most photographic lenses the iris is inside the objective, and it is often adjustable in diameter to control the image illumination and the depth of field. In telescope and microscope systems the cylindrical mount of the objective lens is generally the limiting aperture or iris of the system; its image, formed behind the eyepiece where the observer's eye must be located to see the whole area being observed, called the field, is then the exit pupil.

The pupils of a lens system can be regarded as the common bases of oblique beams passing through the system from all points in an extended object. In most systems, however, the mounts of some of the lens elements cut into the oblique beams and prevent the beams from being perfectly circular, and the pupils are then not fully filled with light. This effect is known as vignetting and leads to a reduction in illumination in the outer parts of the field of view.

A common feature of many optical systems is a relay lens, which may be introduced to invert an image or to extend the length of the system, as in a military periscope. An example of the use of a relay lens is found in the common rifle sight shown diagrammatically in figure. Here the front lens A is the objective, forming an inverted image of the target on the cross wire or reticle at B. The light then proceeds to the relay lens C, which forms a second image, now erect, at D. Beyond this image is the eyepiece E to render the light parallel so that the image may be seen sharply by the observer. Unfortunately, the oblique beam from the objective will usually miss the relay lens, and so a field lens must be inserted at or near the first image B to bend the oblique beams around and redirect them toward the relay lens. The power of the field lens is chosen so that it will form an image of the objective lens aperture on the relay lens aperture. The iris and entrance pupil of this system coincide at the objective; there is an internal pupil at the relay lens, and the exit pupil lies beyond the eyepiece as shown in figure.

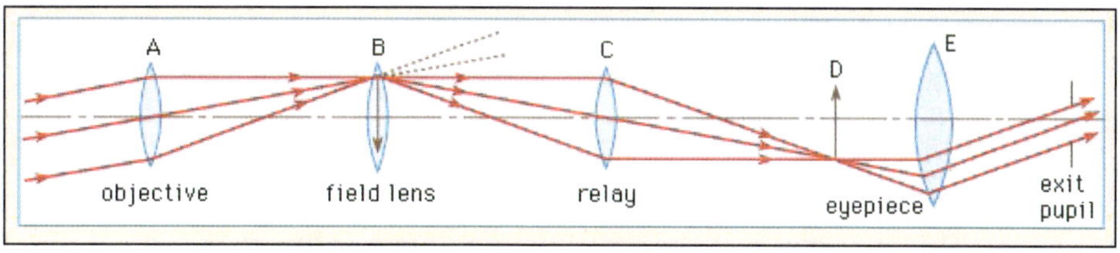

Operating principle of the telescopic rifle sight.

Mirrors

Mirrors are frequently used in optical systems. Plane mirrors may be employed to bend a beam of light in another direction, either for convenience or to yield an image reversed left for right if required. Curved mirrors, concave and convex, may be used in place of lenses as image-forming elements in reflecting telescopes. All of the world's largest telescopes and many small ones are of the reflecting type. Such telescopes use a concave mirror to produce the main image, a small secondary mirror often being added to magnify the image and to place it in a convenient position for observation or photography. Telescope mirrors are commonly made parabolic or hyperbolic in section to correct the aberrations of the image. Originally telescope mirrors were made from polished "speculum metal," an alloy of copper and tin, but in 1856 Justus von Liebig, a German chemist, invented a process for forming a mirror-like layer of silver on polished glass, which was applied to telescope mirrors by the German astronomer C.A. von Steinheil. Today most mirrors are made of glass, coated with either a chemically deposited silver layer or more often one made by depositing vaporized aluminum on the surface. The aluminum surface is as highly reflective as silver and does not tarnish as readily.

A large astronomical mirror presents many problems to the optical engineer, mainly because even a distortion of a few microns of the mirror under its own weight will cause an intolerable blurring of the image. Though many schemes for supporting a mirror without strain have been tried, including one to support it on a bag of compressed air, the problem of completely eliminating mirror distortion remains unsolved. A metal mirror, if well ribbed on the back, may be lighter than a glass mirror and therefore easier to handle, but most metals are slightly flexible and require just as careful support as glass mirrors. Since temperature changes can also cause serious distortion in a mirror, astronomers try to hold observatory temperatures as constant as possible.

Light Sources

Many types of optical instruments form images by natural light, but some, such as microscopes and projectors, require a source of artificial light. Tungsten filament lamps are the most common, but if a very bright source is required, a carbon or xenon arc is employed. For some applications, mercury or other gas discharge tubes are used; a laser beam is often employed in scientific applications. Laser light is brilliant, monochromatic, collimated (the rays are parallel), and coherent (the waves are all in step with each other), any or all of these properties being of value in particular cases.

Detectors

The image formed by an optical system is usually received by the eye, which is a remarkably adaptable and sensitive detector of radiation within the visible region of the electromagnetic spectrum. A photographic film, another widely used detector, has the advantage of yielding a permanent record of events. Since about 1925 many types of electrical detectors of radiation, both within the visible region and beyond it, have been developed. These include photoelectric cells of various kinds in which either a voltage or a resistance is modified by light falling on the device. Many new types of detectors are sensitive far into the infrared spectrum and are used to detect the heat radiated by a flame or other hot object. A number of image intensifiers and converters, particularly for X-ray or infrared radiation, which have appeared since World War II, embody a radiation detector at one end of a vacuum tube and an electron lens inside the tube to relay the image on to a phosphor screen at the other end. This arrangement produces a visible picture that may be observed by eye or photographed to make a permanent record.

Television camera tubes detect real images by electronic scanning, the picture on the viewing tube being a replica of the image in the original camera. The combined application of electronics and optics has become common. An extreme example of electro-optics appears in some space cameras, in which the film is exposed, processed, and then scanned by a tiny point of light; the light passing through the film is picked up by a photocell and transmitted to Earth by radio, where it is made to control the brightness of another point of light scanning a second piece of film in exact synchronism with the scanning spot in the camera. The whole system thus produces a picture on Earth that is an exact replica of the picture photographed in space a few minutes earlier.

Projection Screens

The simplest screen for the projection of slides or motion pictures is, of course, a matte white surface, which may be on a hard base as in outdoor theatres or on a stretched cloth indoors. A theatre screen is often perforated to transmit sound from loudspeakers placed behind it.

Improved screen materials have been developed to increase the brightness of the picture to suit the particular shape of the auditorium. A screen covered with tiny beads tends to send the light back in the general direction of the projector, and is suitable for use at one end of a long, narrow auditorium. Another type of screen is covered with fine embossed vertical grooves; this tends to distribute the light in a horizontal band across the audience with little or no vertical spread. A real advantage of these highly reflective screens is that they tend to reflect ambient room light away from the viewer as by a mirror, so that the pictures appear almost as bright and clear by day as in a darkened room.

Reflecting Prisms

Reflecting prisms are pieces of glass bounded by plane surfaces set at carefully specified angles. Some of these surfaces transmit light, some reflect light, while some serve both functions in succession. A prism is thus an assembly of plane reflectors at relatively fixed angles, which are traversed in succession by a beam of light.

The simplest prism is a triangular block of glass with two faces at right angles and one at an angle of 45°. The face at 45° deflects a beam of light through a right angle. The common Porro prism used in a pair of binoculars contains four 45° reflecting surfaces, two to reverse the beam direction in the vertical plane and two in the horizontal plane. These reflecting faces could be replaced by pieces of mirror mounted on a metal frame, but it is hard to hold mirrors rigidly and harder still to keep them clean. Some microscopes are equipped with a 45° deflection prism behind the eyepiece; this prism may provide two or three reflections depending on the type of image inversion or left-for-right reversal required.

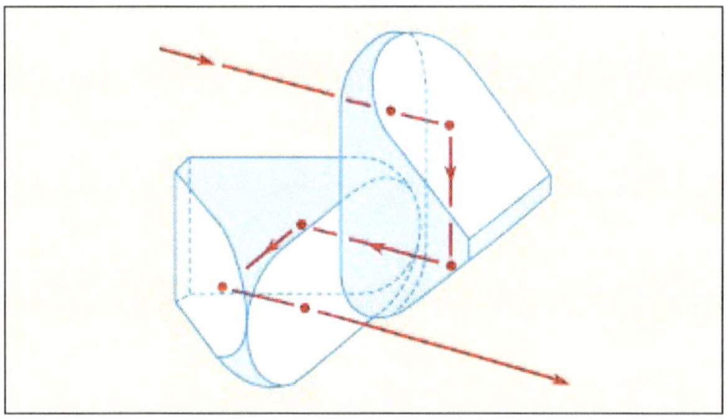

Porro prism.

Prisms containing a semireflecting, semitransmitting surface are known as beam splitters and as such have many uses. An important application is found in some colour television cameras, in which the light from the lens is divided by two beam splitters in succession to form red, green, and blue images on the faces of three image tubes in the camera.

Dispersing Devices

There are two forms of dispersing element used to spread out the constituent colours of a beam of light into a "spectrum," namely a prism and a grating. The prism, known to Newton, is the older; it separates the colours of the spectrum because the refractive index of the glass is lowest for red light and progressively increases through the yellow and green to the blue, where it is highest. Prism spectroscopes and spectrographs are made in a variety of forms and sizes, but in all cases the blue end of the spectrum is greatly spread out while the red end is relatively compressed.

A diffraction grating is a ruled mirror or transparent plate of glass having many thousands of fine parallel grooves to the inch. It separates the colours of the spectrum by a process of diffraction. Each groove diffracts, or scatters, light in all directions, and in the case of light of one particular wavelength, there will be one direction in which the light wave from one groove lags behind the light wave from the next groove by precisely one or more whole wavelengths. This results in a strong beam of diffracted light in that direction and darkness in all other directions. Since each spectral colour corresponds to a different wavelength, the grating spreads out the spectrum into a fan where it can be observed or photographed. The red rays are bent most and the blue rays least, the opposite of the situation with a prism.

Although a prism or grating is the essential dispersing element in a spectrograph, a fine slit and additional lenses or focussing mirrors must be used to form a sharply defined spectrum. Prism spectroscopes are, of course, limited to those wavelengths for which the prism material is transparent; a reflecting grating can be used for any wavelength that the material will reflect.

Filters and Thin Films

A colour filter is a sheet of transparent material that modifies a light beam by selective absorption of some colours in relation to others. A neutral filter absorbs all wavelengths equally and merely serves to reduce the intensity of a beam of light without changing its colour.

Filters may be made from sheets of coloured glass, plastic, or dyed gelatin, and in some cases glass cells filled with liquid have been used. Since World War II, another type of filter depending on the interference of light has been developed in which one or more metallic or other types of films of controlled thickness have been deposited on a glass plate, the layers being so thin as to cause selective interference of some wavelengths in relation to others and thus act as a nonabsorbing filter. In this case the rejected colours are reflected instead of being absorbed.

Polarizing filters have the property of transmitting light that vibrates in one direction while absorbing light that vibrates in a perpendicular direction. These filters are used extensively in scientific instruments. In sunglasses and when placed over a camera lens, polarizing filters reduce unwanted reflections from nonmetallic surfaces. Polarizing spectacles have been used to separate the left-eye and right-eye beams in the projection of stereoscopic pictures or movies.

Fibre-optics Bundles

A thin rod or fibre of glass or other transparent material transmits light by repeated internal reflections, even when the rod is somewhat curved. An ordered bundle of rods or fibres is thus capable of taking an image projected upon one end of the bundle and reproducing it at the other end. A fibre-optics bundle can be fused together into a rigid channel, or it may be left flexible, only the ends being rigidly fastened together. Because a fibre bundle is exceedingly delicate, it must be handled with care; breaking a fibre would cause a black dot to appear in the reproduced image.

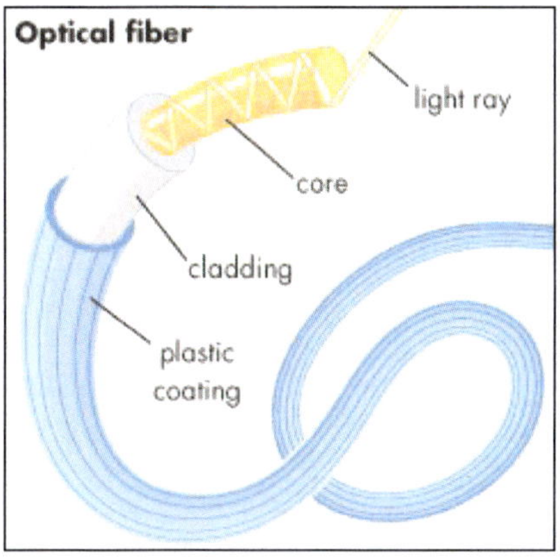

Optical fibre: Light ray passing through an optical fibre.

Nonclassical Imaging Systems

Besides the familiar optical systems cited above, there are many nonclassical optical elements that are used to a limited extent for special purposes. The most familiar of these is the aspheric (nonspherical) surface. Because plane and spherical surfaces are the easiest to generate accurately on glass, most lenses contain only such surfaces. It is occasionally necessary, however, to use some other axially symmetric surface on a lens or mirror, generally to correct a particular aberration. An example is the parabolic surface used for the primary mirror of a large astronomical telescope; another is the elliptic surface molded on the front of the little solid glass reflector units used on highway signs.

Another commonly used optical surface is the side of a cylinder. Such surfaces have power only in the meridian perpendicular to the cylinder axis. Cylindrical lenses are therefore used wherever it is desired to vary the magnification from one meridian to a perpendicular meridian. Cylindrical surfaces are employed in the anamorphic lenses used in some wide-screen motion-picture systems to compress the image horizontally in the camera and stretch it back to its original shape in the projected image.

To correct astigmatism in the eye, many spectacles are made with toric surfaces—i.e., with a stronger curvature in one meridian than in the perpendicular meridian, like the bowl of a teaspoon. These surfaces are generated and polished by special machines and are made by the million every year.

Another nonclassical optical system is the bifocal or trifocal spectacle lens. They are made either by forming two or three separate surfaces on a single piece of glass or obtaining additional power by fusing a piece of high-index glass on to the front of the main lens and then polishing a single spherical surface over both glasses.

Two French scientists, Georges-Louis Buffon and Augustin-Jean Fresnel, in the 18th century suggested forming a lens in concentric rings to save weight, each ring being a portion of what would normally be a continuous spherical surface but flattened out. On a large scale, Fresnel lenses have been used in lighthouses, floodlights, and traffic signals, and as cylindrical ship's lanterns. With fine steps a few thousandths of an inch wide, molded plastic Fresnel lenses are often used as condensers in overhead projectors and in cameras as a field lens in contact with a ground-glass viewing screen.

Lenses have occasionally been made with one surface taking the form of a flattened cone. Such lenses produce a long, linear image of a point source, lying along the lens axis; for this reason they are commonly referred to as axicons. They have been used to produce a straight line of light in space for aligning machines and shafting, but since about 1965 the beam from a gas laser has generally been used instead.

Lens Aberrations

Seidel Sums

If a lens were perfect and the object were a single point of monochromatic light, then, as noted above, the light wave emerging from the lens would be a portion of a sphere centred about the ideal image point, lying in the paraxial image plane at a height above the axis given by the Lagrange theorem. In practice, however, this condition is most unlikely to occur; it is much more probable that the emerging wave will depart slightly from a perfect sphere, the departure varying from point to point over the lens aperture. This departure is extremely small, being of the order of the wavelength of light that is only half a micron, so it would be impossible to show this departure on a drawing. It can be represented mathematically, however, in the following way: The coordinates of a point in the exit-pupil aperture will be represented by x_0 and y_0, the y_0 coordinate lying in the meridian plane containing the object point and the lens axis. The departure of the wave from the ideal sphere is generally called *OPD*, meaning optical path difference. It can be shown that *OPD* is related to x_0 and y_0 by five constants S_1 through S_5, and the quantity h'_0,

$$OPD = S_1\left(x_0^2 + y_0^2\right)^2 + S_2 y_0\left(x_0^2 + y_0^2\right)h'_0 + S_3\left(x_0^2 + 3y_0^2\right)h'^2_0 + S_4\left(x_0^2 + y_0^2\right)h'^2_0 + S_5 y_0 h'^3_0 .$$

Each of these five terms is considered to be a separate "aberration," the coefficients $S_1, \ldots S_5$, being called Seidel sums after the 19th-century German scientist L.P. Seidel, who identified the imperfections. These aberrations are respectively spherical, coma, astigmatism, Petzval field curvature, and distortion. The symbol h'_0 refers to the height of the final image point above the lens axis, and hence it defines the obliquity of the beam.

The five Seidel sums can be calculated by tracing a paraxial ray from object to image through the lens and by tracing also a paraxial principal ray from the centre of the aperture stop outward in both directions toward the object and image, respectively. The angle of incidence *i* and the ray slope angle *u* of each of these paraxial rays at each surface are then listed and inserted into the following expressions for the five sums. The angle u'_0 represents the final emerging slope of the paraxial ray.

The calculation starts by determining the radius *A* of the exit pupil by $A = \sqrt{x_0^2 + y_0^2}$ and also the quantity *K* at each surface by,

$$K = \frac{1}{2} yn\left(\frac{n}{n'} - 1\right)\left(i - u'\right).$$

The corresponding *Kpr* for the paraxial principal ray is also determined at each surface. Then, the five aberrations may be written,

$$S_1 = \frac{1}{4A_4}\sum Ki^2 \qquad S_2 = \frac{1}{A^3 h'_0}\sum Kii_{pr}$$

$$S_3 = \frac{1}{2A^2 h_0'^2}\sum Ki_{pr}^2 \qquad S_4 = \frac{u_0'^2}{4A^2}\sum \frac{n-n'}{nn'r}$$

$$S_5 = \frac{1}{Ah_0'^3}\sum\left[K_{pr} ii_{pr} + \frac{1}{2} h'_0 u'_0 \left(u_{pr}'^2 - u_{pr}^2\right)\right].$$

To interpret these aberrations, the simplest procedure is to find the components x', y' of the displacement of a ray from the Lagrangian image point in the paraxial focal plane, by differentiating the OPD expression given above. The partial derivatives $\partial \mathrm{OPD} / \partial x_0$ and $\partial \mathrm{OPD} / \partial y_0$ represent respectively the components of the slope of the wave relative to the reference sphere at any particular point (x_0, y_0). Hence, because a ray is always perpendicular to the wave, the ray displacements in the focal plane can be found by,

$$x' = f\frac{\partial \mathrm{OPD}}{\partial x_0} \quad \text{and} \quad y' = f\frac{\partial \mathrm{OPD}}{\partial y_0},$$

in which f is the focal length of the lens. The aggregation of rays striking the focal plane will indicate the kind of image that is characteristic of each aberration.

This procedure will be applied to each of the five aberration terms separately, assuming that all the other aberrations are absent. Obviously, in a perfect lens x′ and y′ are zero because OPD is zero. It must be remembered, however, that by using rays instead of waves, all fine-structure effects caused by diffraction will be lost, and only the macroscopic image structure will be retained.

Spherical Aberration

The first term in the OPD expression is OPD $= S_1\left(x_0^2 + y_0^2\right)^2$ Hence,

$$x' = f\frac{\partial \text{OPD}}{\partial x_0} = 4fA^2S_1 \cdot x_0 \quad \text{and} \quad y' = f\frac{\partial \text{OPD}}{\partial y_0}4fA^2S_1 \cdot y_0.$$

These displacements can both be eliminated simultaneously by applying a longitudinal shift L to the focal plane. This changes x' by $-Lx_0/f$ and y' by $-Ly_0/f$; hence, if L is made equal to $4f^2A^2S_1$, both ray displacements vanish. The aberration, therefore, represents a condition in which each zone of the lens has a different focus along the axis, the shift of focus from the paraxial image being proportional to A^2. This is known as spherical aberration.

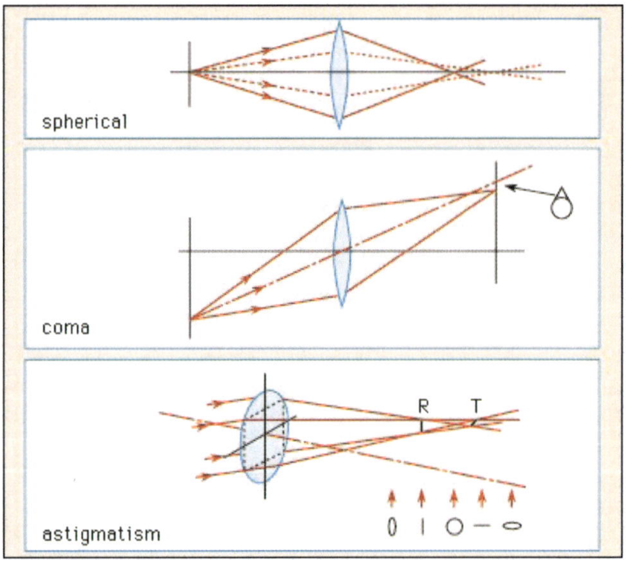

Lens aberrations.

Coma

The S_2 term in the OPD expression represents the aberration called coma, in which the image of a point has the appearance of a comet. The x' and y' components are as follows:

$$x' = fh_0'S_2\left(2x_0y_0\right)$$
$$y' = fh_0'S_2\left(x_0^2 + 3y_0^2\right).$$

When this aberration is present, each circular zone of the lens forms a small ringlike image in the focal plane, the rings formed by successive concentric zones of the lens fitting into two straight envelope lines at 60° to each other. Because the brightness of this image is greatest at the tip, coma tends to form a one-sided haze on images in the outer parts of the field.

Astigmatism

If only the S_3 term is present, then,

$$x' = 2fh_0'^2 S_3 (x_0)$$
$$y' = 2fh_0'^2 S_3 (3y_0).$$

For any one zone of the lens, x' and y' describe a vertical ellipse with major axis three times the minor axis. The images formed by all the smaller zones of the lens fit into this ellipse and fill it out with a uniform intensity of light. If the image plane is moved along the axis by a distance L, as in focussing a camera, then, at $L = 2f^2 h_0'^2 S_3$, the ellipse shrinks to a radial focal line (R). Twice this displacement yields a circle; three times this L gives a tangential focal line (T), which is followed by an ellipse with its major axis in the x direction, as in figure, bottom. The usual effect of astigmatism in an image is the appearance of radial or tangential blurring in the outer parts of the field.

Petzval Curvature

For the S_4 term taken alone,

$$x' = 2fh_0'^2 S_4 \cdot x_0$$
$$y' = 2fh_0'^2 S_4 \cdot y_0$$

The image of a point is now a small circle that contracts to a point at a new focus situated at a longitudinal distance $L = 2f^2 h_0'^2 S_4$ from the paraxial image. As the longitudinal displacement of the focus is proportional to the square of the image height h_0', this aberration represents a pure field curvature without any accompanying loss of definition (all lines remain sharp). It is named after the Hungarian mathematician József Petzval, who studied its properties in the early 1840s. The effect of Petzval curvature can be somewhat offset by the deliberate introduction of sufficient overcorrected astigmatism, as was done in all the pre-anastigmat photographic objectives. This added astigmatism is, of course, undesirable, and in order to design an anastigmat lens having a flat field free from astigmatism, it is necessary to reduce the Petzval sum S_4 drastically.

For a succession of thin lenses (1, 2, 3,...etc.) in a system, the Petzval sum becomes simply $1/f_1n_1 + 1/f_2n_2 + 1/f_3n_3 + \ldots$ etc., in which f is the focal length of each element and n is its refractive index. Therefore, to reduce the sum and minimize this aberration, relatively strong negative elements of low-index glass can be combined with positive elements of high-index glass. The positive and negative elements must be axially separated to provide the lens with a useful amount of positive power. The introduction of high-index barium crown glass with a low dispersive power in the 1880s initiated the development of anastigmat lenses.

Distortion

For the S_5 aberration,

$$x' = 0$$

$$y' = fh_0'^3 S_5.$$

When this aberration is present, the entire image point is displaced toward or away from the axis by an amount proportional to the third power of the transverse distance h_0' of the image from the axis. This leads to the formation of an image of a square that is either a barrel-shaped or a cushion-shaped figure.

It is to be noted that the five Seidel aberrations represent the largest and most conspicuous defects that can arise in an uncorrected optical system. Even in the best lenses in which these five aberrations have been perfectly corrected for one zone of the lens and for one point in the field, however, there will exist small residuals of these aberrations and of many other higher order aberrations also, which are significantly different from the classical types just described. The typical aberration figures shown in figure are, of course, grossly exaggerated, and actually it requires some magnification of a star image to render these appearances clearly visible. Nevertheless, they are important enough to require drastic reduction in high-quality lenses intended to make sharp negatives capable of considerable enlargement.

Image Brightness

General Relations

All photometric concepts are based on the idea of a standard candle, lamps having accurately known candle power being obtainable from the various national standards laboratories. The ratio of the candle power of a source to its area is called the luminance of the source; luminances range from about 2,000 candles per square millimetre at the surface of the Sun down to about 3×10^{-6} candle per square centimetre (3×10^{-6} stilb) for the luminous paint on a watch dial. Ordinary outdoor scenes in daylight have an average luminance of several hundred candles per square foot. The quantity of light flux flowing out from a source is measured in lumens, the lumen being defined as the amount of flux radiated by a small "point" source of one candle power into a cone having a solid angle of one steradian. When light falls upon a surface it produces illumination (i.e., illuminance), the usual measure of illuminance being the foot-candle, which is one lumen falling on each square foot of receiving surface.

It is often important to be able to calculate the brightness of an image formed by an optical system, because photographic emulsions and other light receptors cannot respond satisfactorily if the light level is too low. The problem is to relate the luminance of an object with the illuminance in the image, knowing the transmittance and aperture of the optical system. A small area A of a plane object having a luminance of B candles per square unit will have a normal intensity of AB candles. This source radiates light into a cone of semi-angle U, limited, for example, by the rim of a lens. The light flux (F) entering the cone can be found by integration to be,

$$F = \pi AB \sin^2 U \quad \text{lumens.}$$

If the object luminance is expressed as B_L lamberts, the lambert being an alternative luminance unit equal to 1/π (i.e., 0.32) candle per unit area, the flux (*F*) is,

$$F = AB_L \sin^2 U \quad \text{lumens,}$$

because there are π times as many lamberts in a given luminance as there are candles per unit area.

A fraction *t* of this flux finds its way to the image, *t* being the lens transmittance, generally about 0.8 or 0.9 but less if a mirror is involved. The area of the image is Am^2, in which *m*, the magnification, is given by,

$$m = \frac{\sin U}{\sin U'}.$$

Hence, the image illuminance (E) is,

$$E = \left(t\pi AB \sin^2 U\right) \div \left(A \sin^2 U / \sin^2 U'\right) = t\pi B \sin^2 U' \ \ or$$

$$E = tB_L \sin^2 U'.$$

The image illuminance thus depends only on the luminance of the source and the cone angle of the beam proceeding from the lens to the image. This is a basic and most important relation underlying all calculations of image illuminance.

It is often more convenient to convert the angle U′ into other better known quantities, such as the f-number of the lens and the image magnification. The relation here is,

$$\sin U' = \frac{1}{2(f - \text{number})(1 + m/m_p)}.$$

The *f*-number of the lens is defined as the ratio of the focal length to the diameter of the entrance pupil; *m* is the image magnification; and *mp* is the pupil magnification—i.e., the diameter of the exit pupil divided by the diameter of the entrance pupil. Combining equations $E = \left(t\pi AB \sin^2 U\right) \div \left(A \sin^2 U / \sin^2 U'\right) = t\pi B \sin^2 U'$ and $\sin U' = \frac{1}{2(f - \text{number})(1 + m/m_p)}$ gives,

$$\text{Image illuminance} = E \frac{\pi t B}{4(f - \text{number})^2 (1 + m/m_p)^2}.$$

As an example in the use of this relation, if it is supposed that an f/2 lens is being used to project an image of a cathode-ray tube at five times magnification, the tube luminance being 5,000 foot-lamberts (1.7 candles per square centimetre), the lens transmittance is 0.8, and the pupil magnification is unity. Then the image illuminance will be,

$$E = \frac{0.8 \times 5,000}{4 \times 4 \times 36} = 6.9 \text{ foot-candles.}$$

The image is very much less bright than the object, a fact that becomes clear to anyone attempting to provide a bright projected image in a large auditorium.

Distribution of Illumination over an Image

So far only the illumination at the centre of an image has been considered, but the distribution of illumination over a wide field is often important. In the absence of any lens, the small plane source already considered radiates in a direction inclined at an angle ϕ to the axis with an intensity $AB \cos \phi$. This light has to travel farther than the axial light to reach a screen, and then it strikes the screen at another angle ϕ. The net result is that the oblique illumination on the screen is smaller than the axial illumination by the factor $\cos^4 \phi$.

The same law can be applied to determine the oblique illumination due to a lens, assuming a uniform extended diffusing source of light on the other side of the lens. In this case, however, the exit pupil will not in general be a perfect circle because of possible distortion of the iris by that part of the optical system lying between the iris and the image. Also, any mechanical vignetting in the lens will make the aperture noncircular and reduce still further the oblique illumination. In a camera this reduction in oblique illumination results in darkened corners of the picture, but, if the reduction in brightness is gradual, it is not likely to be detected because the eye adapts quickly to changing brightness as the eyes scan over the picture area. Indeed, a 50 percent drop in brightness between the centre and corners of an ordinary picture is scarcely detectable.

Visual Brightness

The apparent brightness of things seen by the eye follows the same laws as any other imaging system, because the apparent brightness is measured by the illuminance in the image that is projected on the retina. The angle U' in equation $E = \left(t\pi AB \sin^2 U\right) \div \left(A \sin^2 U / \sin^2 U'\right) = t\pi B \sin^2 U'$ inside the eye is determined by the size of thc pupil of the eye, which varies from about one millimetre to about eight millimetres, depending on the brightness of the environment. Apart from this variation, retinal illuminance is directly proportional to object luminance, and objects having the same luminance appear equally bright, no matter at what distance they are observed.

From this argument, it is clear that no visual instrument, such as a telescope, can possibly make anything appear brighter than when viewed directly. To be sure, a telescope having a large objective lens accepts more light from an object in proportion to the area of the lens aperture, but it magnifies the image area in the same proportion; so the increased light is spread over an increased area of the retina, and the illuminance remains unchanged. Actually, the telescopic view is always dimmer than the direct view because of light losses in the telescope due to glass absorption and surface reflections and because the exit pupil of the telescope may be smaller than the pupil of the eye, thus reducing the angle U'.

The case of a star being observed through a telescope is quite different, because no degree of magnification can possibly make a star appear as anything other than a point of light. Hence, star images appear brighter in proportion to the area of the telescope objective (assuming that the exit pupil is larger than the eye pupil), and the visibility of a star against the sky background is thus improved in proportion to the square of the diameter of the telescope objective lens.

Optics and Information Theory

General Observations

A new era in optics commenced in the early 1950s following the impact of certain branches of electrical engineering—most notably communication and information theory. This impetus was sustained by the development of the laser in the 1960s.

The initial tie between optics and communication theory came because of the numerous analogies that exist between the two subjects and because of the similar mathematical techniques employed to formally describe the behaviour of electrical circuits and optical systems. A topic of considerable concern since the invention of the lens as an optical imaging device has always been the description of the optical system that forms the image; information about the object is relayed and presented as an image. Clearly, the optical system can be considered a communication channel and can be analyzed as such. There is a linear relationship (i.e., direct proportionality) between the intensity distribution in the image plane and that existing in the object, when the object is illuminated with incoherent light (e.g., sunlight or light from a large thermal source). Hence, the linear theory developed for the description of electronic systems can be applied to optical image-forming systems. For example, an electronic circuit can be characterized by its impulse response—that is, its output for a brief impulse input of current or voltage. Analogously, an optical system can be characterized by an impulse response that for an incoherent imaging system is the intensity distribution in the image of a point source of light; the optical impulse is a spatial rather than a temporal impulse—otherwise the concept is the same. Once the appropriate impulse response function is known, the output of that system for any object intensity distribution can be determined by a linear superposition of impulse responses suitably weighted by the value of the intensity at each point in the object. For a continuous object intensity distribution this sum becomes an integral. While this example has been given in terms of an optical imaging system, which is certainly the most common use of optical elements, the concept can be used independent of whether the receiving plane is an image plane or not. Hence, for example, an impulse response can be defined for an optical system that is deliberately defocussed or for systems used for the display of Fresnel or Fraunhofer diffraction patterns. (Fraunhofer diffraction occurs when the light source and diffraction patterns are effectively at infinite distances from the diffracting system, and Fresnel diffraction occurs when one or both of the distances are finite).

Temporal Frequency Response

A fundamentally related but different method of describing the performance of an electronic circuit is by means of its temporal frequency response. A plot is made of the response for a series of input signals of a variety of frequencies. The response is measured as the ratio of the amplitude of the signal obtained out of the system to that put in. If there is no loss in the system, then the frequency response is unity (one) for that frequency; if a particular frequency fails to pass through the system, then the response is zero. Again, analogously the optical system may also be described by defining a spatial frequency response. The object, then, to be imaged by the optical system consists of a spatial distribution of intensity of a single spatial frequency—an object the intensity of which varies as $(1 + a \cos \omega x)$, in which x is the spatial coordinate, a is a constant called the contrast, and ω is a variable that determines the physical spacing of the peaks in the intensity distribution. The image is recorded for a fixed value of a and ω and the contrast in the image measured. The ratio of this contrast to a is the response for this particular spatial frequency defined by ω. Now if ω is varied and the measurement is repeated, a frequency response is then obtained.

Nonlinear Optical Systems

Many optical systems are nonlinear, just as many electronic systems are nonlinear. Photographic film is a nonlinear optical element in that equal increments of light energy reaching the film do not always produce equal increments of density on the film.

A different type of nonlinearity occurs in image formation. When an object such as two stars is imaged, the resultant intensity distribution in the image is determined by first finding the intensity distribution formed by each star. These distributions must then be added together in regions where they overlap to give the final intensity distribution that is the image. This example is typical of an incoherent imaging system—i.e., the light emanating from the two stars is completely uncorrelated. This occurs because there is no fixed phase relationship between the light emanating from the two stars over any finite time interval.

A similar nonlinearity arises in objects illuminated by light from the Sun or other thermal light source. Illumination of this kind, when there is no fixed relationship between the phase of the light at any pair of points in the incident beam, is said to be incoherent illumination. If the illumination of the object is coherent, however, then there is a fixed relationship between the phase of the light at all pairs of points in the incident beam. To determine the resultant image intensity under this condition for a two point object requires that the amplitude and phase of the light in the image of each point be determined. The resultant amplitude and phase is then found by summation in regions of overlap. The square of this resultant amplitude is the intensity distribution in the image. Such a system is nonlinear. The mathematics of nonlinear systems was developed as a branch of communication theory, but many of the results can be used to describe nonlinear optical systems.

This new description of optical systems was extremely important to, but would not alone account for, the resurgence of optical research and development. This new approach resulted in the development of whole new branches of study, including optical processing and holography. It also had an effect, together with the development of digital computers, on the concepts and versatility of lens design and testing. Finally, the invention of the laser, a device that produces coherent radiation, and the development and implementation of the theory of partially coherent light gave the added impetus necessary to change traditional optics into a radically new and exciting subject.

Image Formation

Impulse Response

An optical system that employs incoherent illumination of the object can usually be regarded as a linear system in intensity. A system is linear if the addition of inputs produces an addition of corresponding outputs. For ease of analysis, systems are often considered stationary (or invariant). This property implies that if the location of the input is changed, then the only effect is to change the location of the output but not its actual distribution. With these concepts it is then only necessary to find an expression for the image of a point input to develop a theory of image formation. The intensity distribution in the image of a point object can be determined by solving the equation relating to the diffraction of light as it propagates from the point object to the lens, through the lens, and then finally to the image plane. The result of this process is that the image intensity is the intensity in the Fraunhofer diffraction pattern of the lens aperture function (that is, the square of the Fourier transform of the lens aperture function; a Fourier transform is an integral equation involving periodic components). This intensity distribution is the intensity impulse response (sometimes called point spread function) of the optical system and fully characterizes that optical system.

With the knowledge of the impulse response, the image of a known object intensity distribution can be calculated. If the object consists of two points, then in the image plane the intensity impulse response function must be located at the image points and then a sum of these intensity distributions made. The sum is the final image intensity. If the two points are closer together than the half width of the impulse response, they will not be resolved. For an object consisting of an array of isolated points, a similar procedure is followed—each impulse response is, of course, multiplied by a constant equal to the value of the intensity of the appropriate point object. Normally, an object will consist of a continuous distribution of intensity, and, instead of a simple sum, a convolution integral results.

Transfer Function

The concept of the transfer function of an optical system can be approached in several ways. Formally and fundamentally it is the Fourier transform of the intensity impulse response. Because the impulse response is related to the lens aperture function, so is the transfer function. In particular, the transfer function can be obtained from a knowledge of the aperture function by taking the function and plotting the resultant overlapping areas as the aperture function is slid over itself (i.e., the autocorrelation of the aperture function).

Conceptually, however, the transfer function is best understood by considering the object intensity distribution to be a linear sum of cosine functions of the form $(1 + a\cos 2\pi\mu x)$, in which a is the amplitude of each component of spatial frequency μ. The image of a cosine intensity distribution is a cosine of the same frequency; only the contrast and phase of the cosine can be affected by a linear system. The image of the above object intensity distribution can be represented by, in which b is the amplitude of the output cosine of frequency μ and ϕ is the phase shift. The transfer function, τ(μ), for that frequency is then given by the ratio of the amplitudes:

$$\tau(u) = \frac{b}{a} e^{j\varphi(\mu)}.$$

If μ is now varied, the spatial frequency response of the system is measured by determining τ(μ) for the various values of μ. It should be noted that τ(μ) is in general complex (containing a term with $\sqrt{-1}$).

The transfer function, like the impulse response, fully characterizes the optical system. To make use of the transfer function to determine the image of a given object requires that the object be decomposed into a series of periodic components called its spatial frequency spectrum. Each term in this series must then be multiplied by the appropriate value of the transfer function to determine the individual components of the series that is the spatial frequency spectrum of the image—a transformation of this series will give the image intensity. Thus, any components in the object spectrum that have a frequency for which τ(μ) is zero will be eliminated from the image.

Partially Coherent Light

Development and Examples of the Theory

Image formation is concerned above with incoherent object illumination, which results in an image formed by the addition of intensities. The study of diffraction and interference, on the other hand, requires coherent illumination of the diffracting object, the resulting diffracted optical field being determined by an addition of complex amplitudes of the wave disturbances. Thus, two different mechanisms exist for the addition of light beams, depending upon whether the beams are coherent or incoherent with respect to each other. Unfortunately, this is not the whole story; it is not sufficient to consider only the two situations of strictly coherent and strictly incoherent light. In fact, strictly incoherent fields are only approximately obtainable in practice. Furthermore, the possibility of intermediate states of coherence cannot be ignored; it is necessary to describe the result of mixing incoherent light with coherent light. It was to answer the question How coherent is a beam of light? (or the equivalent one, How incoherent is a beam of light?) that the theory of partial coherence was developed. Marcel Verdet, a French physicist, realized in the 19th century that even sunlight is not completely incoherent, and two objects separated by distances of over approximately $^1/_{20}$ millimetre will produce interference effects. The eye, operating unaided in sunlight, does not resolve this separation distance and hence can be considered to be receiving an incoherent field. Two physicists, Armand Fizeau in France and Albert Michelson in the United States, were also aware that the optical field produced by a star is not completely incoherent, and hence they were able to design interferometers to measure the diameter of stars from a measurement of the partial coherence of the starlight. These early workers did not think in terms of partially coherent light, however, but derived their results by an integration over the source. At the other extreme, the output from a laser can produce a highly coherent field.

The concepts of partially coherent light can best be understood by means of some simple experiments. A circular uniform distant source produces illumination on the front of an opaque screen containing two small circular apertures, the separation of which can be varied. A lens is located behind this screen, and the resultant intensity distribution in its focal plane is obtained. With either aperture open alone, the intensity distribution observed is such that it is readily associated with the diffraction pattern of the aperture, and it may thus be concluded that the field is coherent over the dimensions of the aperture. When the two apertures are opened together and are at their closest separation, two-beam interference fringes are observed that are formed by the division of the incident wave front by the two apertures. As the separation of the apertures increases, the observed interference fringes get weaker and finally disappear, only to reappear faintly as the separation is further increased. As the separation of the apertures is increased, these results show that the fringe spacing decreases; the intensities of the fringe minima are never zero; the relative intensity of the maxima above the minima steadily decreases; the absolute value of the intensity of the maxima decreases and that of the minima increases; eventually, the fringes disappear, at which point the resultant intensity is just twice the intensity observed with one aperture alone (essentially an incoherent addition); the fringes reappear with a further increase in separation of the aperture, but the fringes contain a central minimum, not a central maximum.

If the intensities of the two apertures are equal, then the results can be summarized by defining a quantity in terms of the maximum intensity (*Imax*) and the minimum intensity

(I_{min}), called the visibility (*V*) of the fringes—*i.e.*, $V = (I_{max} - I_{min})/(I_{max} + I_{min})$. The maximum value of the visibility is unity, for which the light passing through one aperture is coherent with respect to the light passing through the other aperture; when the visibility is zero, the light passing through one aperture is incoherent with respect to the light passing through the other aperture. For intermediate values of *V* the light is said to be partially coherent. The visibility is not a completely satisfactory description because it is, by definition, a positive quantity and cannot, therefore, include a description of item above. Furthermore, it can be shown by a related experiment that the visibility of the fringes can be varied by adding an extra optical path between the two interfering beams.

The Mutual Coherence Function

The key function in the theory of partially coherent light is the mutual coherence function $\Gamma_{12}(\mathrm{T}) = \Gamma(x_1, x_2, \mathrm{T})$, a complex quantity, which is the time averaged value of the cross correlation function of the light at the two aperture points x_1 and x_2 with a time delay T (relating to a path difference to the point of observation of the interference fringes). The function can be normalized (*i.e.*, its absolute value set equal to unity at T = 0 and x_1 x_2) by dividing by the square root of the product of the intensities at the points x_1 and x_2 to give the complex degree of coherence, hence

$$\gamma_{12}(\tau) = \frac{\Gamma_{12}(\tau)}{\sqrt{I(x_1)I(x_2)}}.$$

The modulus of $\gamma_{12}(\mathrm{T})$ has a maximum value of unity and a minimum value of zero. The visibility defined earlier is identical to the modulus of the complex degree of coherence if $I(x_1) = I(x_2)$.

Often the optical field can be considered to be quasimonochromatic (approximately monochromatic), and then the time delay can be set equal to zero in the above expression, thus defining the mutual intensity function. It is often convenient to describe an optical field in terms of its spatial and temporal coherence by artificially separating out the space- and time-dependent parts of the coherence function. Temporal coherence effects arise from the finite spectral width of the source radiation; a coherence time Δt can be defined as $1/\Delta \nu$, in which $\Delta \nu$ is the frequency bandwidth. A related coherence length Δl can also be defined as $c / \Delta \nu = \lambda^2 / \Delta \lambda^2$, in which *c* is the velocity of light, λ is the wavelength, and $\Delta\lambda$ the wavelength bandwidth. Providing that the path differences in the beams to be added are less than this characteristic length, the beams will interfere.

The term spatial coherence is used to describe partial coherence arising from the finite size of an incoherent source. Hence, for the equipath position for the addition of two beams, a coherence interval is defined as the separation of two points such that the absolute value $|\gamma_{12}(0)|$ is some prechosen value, usually zero.

The mutual coherence function is an observable quantity that can be related to the intensity of the field. The partially coherent field can be propagated by use of the mutual coherence function in a similar way to the solution of diffraction problems by propagation of the complex amplitude. The effects of partially coherent fields are clearly of importance in the description of normally coherent phenomena, such as diffraction and interference, but also in the analysis of normally incoherent phenomena, such as image formation. It is notable that image formation in coherent light is not linear in intensity but is linear in the complex amplitude of the field, and in partially coherent light the process is linear in the mutual coherence.

Optical Processing

Coherent Optical Systems

Optical processing, information processing, signal processing, and pattern recognition are all names that relate to the process of spatial frequency filtering in a coherent imaging system—specifically, a method in which the Fraunhofer diffraction pattern (equivalently the spatial frequency spectrum or the Fourier transform) of a given input is produced optically and then operated upon to change the information content of the optical image of that input in a predetermined way.

The idea of using coherent optical systems to allow for the manipulation of the information content of the image is not entirely new. The basic ideas are essentially included in Abbe's theory of vision in a microscope first published in 1873; the subsequent illustrative experiments of this theory, notably by Albert B. Porter in 1906, are certainly simple examples of optical processing.

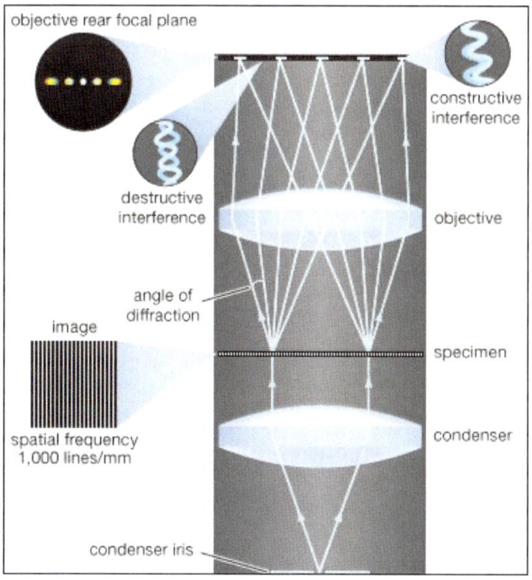

Image formation in a microscope, according to the Abbe theory. Specimens are illuminated by light from a condenser. This light is diffracted by the details in the object plane: the smaller the detailed structure of the object, the wider the angle of diffraction. The structure of the object can be represented as a sum of sinusoidal components. The rapidity of variation in space of the components is defined by the period of each component, or the distance between adjacent peaks in the sinusoidal function. The spatial frequency is the reciprocal of the period. The finer the details, the higher the required spatial frequency of the components that represent the object detail. Each spatial frequency component in the object produces diffraction at a specific angle dependent upon the wavelength of light. Here, for example, a specimen with structure that has a spatial frequency of 1,000 lines per millimetre produces diffraction with an angle of 33.6°. The microscope objective collects these diffracted waves and directs them to the focal plane, where interference between the diffracted waves produces an image of the object.

Abbe's ideas can be interpreted as a realization that image formation in a microscope is more correctly described as a coherent image-forming process than as the more familiar incoherent process. Thus, the coherent light illuminating the object on the microscope stage would be diffracted by that object. To form an image, this diffracted light must be collected by the objective lens of the microscope, and the nature of the image and the resolution would be affected by how much of the diffracted light is collected. As an example, an object may be considered consisting of a periodic variation in amplitude transmittance—the light diffracted by this object will exist in a series of discrete directions (or orders of diffraction). This series of orders contains a zero order propagating along the optical axis and a symmetric set of orders on both sides of this zero order. Abbe correctly discerned what would happen as the microscope objective accepted different combinations of these orders. For example, if the zero order and one first order are collected, then the information obtained will be that the object consisted of a periodic distribution, but the spatial location of the periodic structure is not correctly ascertained. If the other first order of diffracted light is included, the correct spatial location of the periodic structure is also obtained. As more orders are included, the image more closely resembles the object.

Coherent optical data processing became a serious subject for study in the 1950s, partly because of the work of a French physicist, Pierre-Michel Duffieux, on the Fourier integral and its application to optics, and the subsequent use of communication theory in optical research. The work was initiated in France by André Maréchal and Paul Croce, and today a variety of problems can be attempted by the technique. These include removal of raster lines (as in a TV picture) and halftone dots (as in newspaper illustration); contrast enhancement; edge sharpening; enhancement of a periodic or isolated signal in the presence of additive noise; aberration balancing in which a recorded aberrated image can be somewhat improved; spectrum analysis; cross correlation of data; matched and inverse filtering in which a bright spot of light in the image indicates the presence of a particular object.

Filtering

The basic system required for coherent optical processing consists of two lenses. A collimated beam of coherent light is used to transilluminate the object. The first lens produces the characteristic Fraunhofer diffraction pattern of the object, which is the spatial frequency distribution associated with the object. (Mathematically, it is the Fourier transform of the object amplitude distribution.) A filter that consists of amplitude (density) or phase (optical path) variations, or both, is placed in the plane of the diffraction pattern. The light passing through this filter is used to form an image, this step being accomplished by the second lens. The filter has the effect of changing the nature of the image by altering the spatial frequency spectrum in a controlled way so as to enhance certain aspects of the object information. Maréchal gave the descriptive title double diffraction to this type of two-lens system.

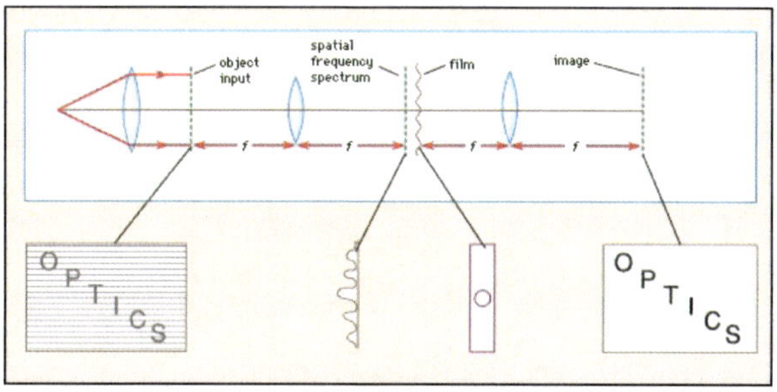

Figure: Two-lens coherent optical processing system, showing how the raster periodicity is removed but the scene information is retained.

The filters can be conveniently grouped into a variety of types depending upon their action. Blocking filters have regions of complete transparency and other regions of complete opacity. The opaque areas completely remove certain portions of the spatial frequency spectrum of the object. The removal of raster lines and halftone dots is accomplished with this type of filter. The object can be considered as a periodic function the envelope of which is the scene or picture—or equivalently the periodic function samples the picture. The diffraction pattern consists of a periodic distribution with a periodicity reciprocally related to the raster periodicity. Centred at each of these periodic locations is the diffraction pattern of the scene. Hence, if the filter is an aperture centred at one of these locations so that only one of the periodic elements is allowed to pass, then the raster periodicity is removed, but the scene information is retained. The problem of the removal of halftone dots is the two-dimensional equivalent of the above process. Because the two-dimensional spatial frequency spectrum of an object is displayed in a coherent optical processing system, it is possible to separate out information by means of its orientation. Other applications of blocking filters include band-pass filters, which again have a direct relationship to the band-pass filters in electronic circuits.

A second type of filter is an amplitude filter that will consist of a continuous density variation. These filters can be produced to achieve the enhancement of contrast of the object input or the differentiation of the object. They are often constructed by controlled exposure of photographic film or evaporation of metal onto a transparent substrate.

Certain optical processing techniques require that the phase of the optical field be changed, and, hence, a filter with no absorption but varying optical thickness is required. Usually, both the amplitude and the phase have to be modified, however, thus requiring a complex filter. In simple cases the amplitude and phase portions can be made separately, the phase filter being manufactured by using an evaporated layer of transparent material, such as magnesium fluoride. Current practice is to fabricate the complex filter by an interferometric method in which the required complex amplitude function is recorded as a hologram.

The phase-contrast microscope can be considered to be an example of an optical processing system, and the concepts understood by reference to figure. Only the simplest form will be considered here. The spatial frequency spectrum of the phase object is formed and the phase of the central portion of that spectrum changed by $\pi/2$ or $3\pi/2$ to produce positive or negative phase contrast, respectively. To improve the contrast of the image an additional filter covering the same area as the phase filter is used that is partially absorbing (*i.e.,* an amplitude filter). The restriction on this process is that the variations of the phase $\phi(x)$ are small so that $e^{i}\phi^{(x)} \cong 1+i\phi(x)$. With incoherent light, phase information is not visible, but many biological samples consist only of variations of refractive index, which results in optical path and hence phase, differences. The image in the phase-contrast microscope is such that the intensity in that image relates linearly to, and hence is a display of, the phase information in the object—*e.g.,* $I(x) \propto 1 \pm 2\phi(x)$ for positive and negative phase contrast, respectively.

One of the important motivations for the study of optical processing methods is to achieve some correction of aberrated images. Considerable technological advantage can be gained if photographs taken with an aberrated optical system in incoherent light can be corrected by subsequent processing. Within definable limits this can be accomplished, but the impulse response or the transfer function of the aberrated system must be known. The recorded image intensity distribution is the convolution of the object intensity with the intensity impulse response of the aberrated system. This record is the input to the coherent optical processing system; the diffraction pattern formed in this system is the product of the spatial frequency spectrum of the object and the transfer function of the aberrated system. Conceptually, the filter has to be the inverse of the transfer function in order to balance out its effect. The final image would then ideally be an image of the object intensity distribution. It is critical, however, that the transfer function has a finite value over only a limited frequency range, and only those frequencies that are recorded by the original aberrated system can be present in the processed image. Hence, for these spatial frequencies that were recorded, some processing can be carried out to get a flatter effective transfer function; both the contrast and the phase of the spatial frequency spectrum may have to be changed because the transfer function is, in general, a complex function. Prime examples are for images aberrated by astigmatism, defocussing, or image motion.

Holography

Holography is a two-step coherent image-forming process in which an intermediate record is made of the complex optical field associated with the object. The invention of the wave-front reconstruction process (now called holography) was first described in 1948 by Dennis Gabor, a Hungarian-born physicist, with a specific application in mind—to attempt to improve the resolution of images formed with electron beams. The technique has, however, had most of its success to date when light beams are employed particularly in the visible part of the spectrum. The first step in the process is to record (often on high-resolution film) the interference pattern produced by the interaction of the light diffracted by the object of interest and a coherent background or reference wave. In the second step, this record, which is the hologram, is illuminated coherently to form an image of the original object. In fact, two images are usually formed—a real image (often called the conjugate image) and a virtual image (often called the primary image). There are two basic concepts that underlie this process: first, the addition of a coherent background (or reference) beam. Two optical fields may be considered, the complex amplitudes of which vary as the cosine of an angle proportional to the space coordinate and as the modulus (absolute magnitude) of the cosine of the angle, respectively. From a measurement of the intensity of these fields it is impossible to distinguish them because both vary as the cosine squared of the space coordinate. If a second coherent optical field is added to each of these two fields, however, then the resultant fields become $(1 + \cos x)$ and $(1 + |\cos x|)$, respectively. The measured intensities are now different, and the actual fields can be determined by taking the square root of the intensity. The amplitude transmittance of a photographic record is, in fact, the square root of the original intensity distribution that exposed the film. In a more general sense, an optical field of the form $a(x)\exp\left[i\phi_1(x)\right]$, in which $a(x)$ is the amplitude and $\phi_1(x)$ is the phase, can be distinguished from a field $a(x)\exp\left[i\phi_2(x)\right]$ by adding a coherent background; the phases $\phi_1(x)$ and $\phi_2(x)$ are then contained as cosine variations of intensity in the resulting pattern. Hence, the problem of recording the phase information of the optical field is circumvented. When the hologram is illuminated, however, the optical field that originally existed in that plane is recreated. To apply the second basic concept—that of an image-forming property—it is necessary to determine what the hologram of a point object is—in actuality it is a sine-wave zone plate or zone lens. If a collimated beam of light is used to illuminate a zone lens, then two beams are produced; the first comes to a real focus, and the other is a divergent beam that appears to have come from a virtual focus. (By comparison, the more classical zone plate has a multitude of real and virtual focuses, and a real lens has but one.) When the object is other than a point, the zone lens is modified by the diffraction pattern of the object; *i.e.*, each point on the object produces its own zone lens, and the resultant hologram is a summation of such zone lenses.

In Gabor's original system the hologram was a record of the interference between the light diffracted by the object and a collinear background. This automatically restricts the process to that class of objects that have considerable areas that are transparent. When the hologram is used to form an image, twin images are formed, as illustrated in figure. The light associated with these images is propagating in the same direction, and hence in the plane of one image light from the other image appears as an out-of-focus component. This type of hologram is usually referred to as an in-line Fresnel hologram because it is the pattern of the object that interferes with the collinear coherent background. The deleterious effects of the second image can be minimized if the hologram is made in the far field of the object so that it is a Fraunhofer diffraction pattern of the object that is involved. This latter technique has found significant application in microscopy, particularly in the measurement of small particles, and in electron microscopy.

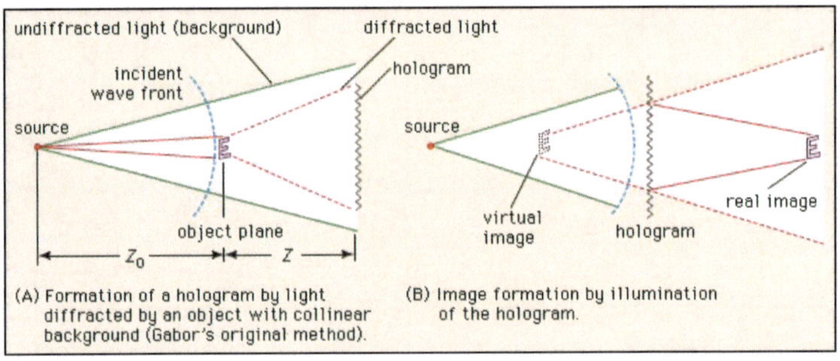

(A) Formation of a hologram by light diffracted by an object with collinear background (Gabor's original method).

(B) Image formation by illumination of the hologram.

Holography.

A more versatile method of recording the hologram is to add a second beam of light as a reference wave to produce the hologram. The hologram is now the record of the interference pattern produced by the light diffracted by the object and this separate reference wave. The reference wave is usually introduced at an angle to the diffracted beam, hence this method is often called off-axis (or side-band) holography. When the hologram is illuminated, the image-forming beams do not propagate in the same direction but are inclined to each other with an angle twice that between the diffracted beam and the original reference beam. Hence, the light associated with an image is completely separated from the other image.

A further technique that has some value and relates to the earlier discussion of optical processing is the production of the so-called generalized or Fourier transform hologram. Here the reference beam is added coherently to a Fraunhofer diffraction pattern of the object or formed by a lens.

The process described so far has been in terms of transmitted light through the object. The methods involving the separate reference beam can be used in reflected light, and the virtual (primary) image produced from the hologram has all the properties of an ordinary image in terms of three-dimensionality and parallax. Normally, a recorded image is only a two-dimensional representation of the object. Full-colour holograms can be recorded by essentially recording three holograms simultaneously—one in red light, one in blue, and one in green.

Applications

Image-forming

The applications mentioned here are in three groups: image-forming applications, non-image-forming applications, and the hologram as an optical element. It is notable that all three groups relate to the basic use of the process rather than specific holographic techniques. The first group involves those applications using image formation when, for a variety of reasons, normal incoherent or coherent image formation is not satisfactory. It is not sufficient merely to replace a normal image process by a holographic technique unless there is some significant gain—i.e., the required record can be obtained more easily or more accurately. Applications that fall into this category are holographic microscopy; particle-size analysis; high-speed photography of various types, particularly of gas flows; data storage and retrieval, including displays; image formation through a random medium; and non-optical holography, particularly acoustic holography.

Non-image-forming

The second group of interest involves those applications that are not image-forming. One of the very real and exciting applications of holography is to the nondestructive testing of fabricated materials. An interesting example of this method is for the testing of tires for the detection of flaws (debonds) that exist between the plies of the tire. The realm of interferometry is thus extended to whole new classes of objects. In a similar but separate development, interference microscopy has been used successfully.

Optical Elements

The third and final group involves those applications that use the hologram as an optical element in its own right. This includes the building of accurate, specialized gratings and the application of holographic filters in coherent optical data processing.

Holography has been adapted to the conventional microscope, which is modified by the inclusion of a separate reference beam so that the light diffracted by the object in the microscope is made to interfere with the light from the reference beam. An increase in the depth of field available is achieved by this type of recording process. The image is produced when the hologram is illuminated again by a coherent beam.

The application of holography to particle-size analysis (e.g., to determine the size distribution of dust and liquid droplets) was really the first of the modern-day applications. In a sense, this, too, can be thought of as microscopy. The principles of Fraunhofer holography were developed to solve this particular problem. Because the particles are in motion, a hologram must be made instantaneously. A pulsed-ruby laser technique is therefore used. The hologram is formed between the light diffracted by the particles or droplets and the coherent background light that passes directly through the sample. In reconstruction, a series of stationary images are formed that can be examined at leisure. Hence, a transient event has been transformed into a stationary image for evaluation.

Data storage and retrieval is perhaps one of the more important applications of holography, which is in the process of development and refinement. Because the information about the image is not localized, it cannot be affected by scratches or dust particles.

Recent advances in materials, particularly those that might be erasable and reusable, have added further interest in holographic optical memories.

Among the non-image-forming applications are interferometry, interference microscopy, and optical processing. Holographic interferometry can be done in several ways. The basic technique involves recording a hologram of the object of interest and then interfering the image produced from this hologram with the coherently illuminated object itself. A variation on this technique would be to form two holograms at different times of the same object as it undergoes testing. The two holograms can then be used together to form two images, which would again interfere. The interference fringes seen would be related to the changes in the object between the two exposures. A third technique uses a time-average hologram, which is particularly applicable to the study of vibrating objects.

There are two applications that come under the heading holographic optical elements—the use of holographic gratings and the use of holographic filters for coherent optical data processing.

Nonlinear Optics

Nonlinear effects in optics are now quite readily observable using the highly coherent and highly energetic laser beams. These effects occur when the output of a system is not linearly related to the input (e.g., a nonlinear electronic amplifier can be built with a gain that increases with signal intensity). The most important nonlinear effect is probably frequency doubling. Optical radiation of a given frequency is propagated through a crystalline material and interacts with that material to produce an output of a different frequency that is twice the input frequency. For example, the 10,600-angstrom infrared output of a neodymium laser can, under suitable conditions, be converted into green light at 5300 angstroms in a crystal of barium strontium niobate.

2

Interference of Light

Interference of light is defined as the phenomenon of reinforcing of two light waves in order to form a greater wave. Fresnel biprism, Lloyd's mirror arrangement, Newton's rings, multiple-beam interferometry, coherence, Fourier transform spectroscopy, etc. are a few of the concepts that fall under its domain. All these concepts of interference of light have been carefully analyzed in the chapter.

Interference is the phenomenon that occurs when two waves meet while traveling along the same medium. The interference of waves causes the medium to take on a shape that results from the net effect of the two individual waves upon the particles of the medium. To begin our exploration of wave interference, consider two pulses of the same amplitude traveling in different directions along the same medium. Let's suppose that each displaced upward 1 unit at its crest and has the shape of a sine wave. As the sine pulses move towards each other, there will eventually be a moment in time when they are completely overlapped. At that moment, the resulting shape of the medium would be an upward displaced sine pulse with an amplitude of 2 units. The diagrams below depict the before and during interference snapshots of the medium for two such pulses. The individual sine pulses are drawn in red and blue and the resulting displacement of the medium is drawn in green.

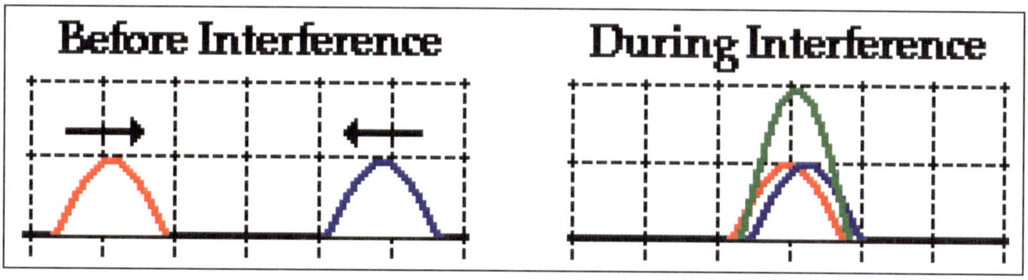

Constructive Interference

Constructive interference is a type of interference that occurs at any location along the medium where the two interfering waves have a displacement in the same direction. In this case, both waves have an upward displacement; consequently, the medium has an upward displacement that is greater than the displacement of the two interfering pulses. Constructive interference is observed at any location where the two interfering waves are displaced upward. But it is also observed when both interfering waves are displaced downward. This is shown in the diagram below for two downward displaced pulses.

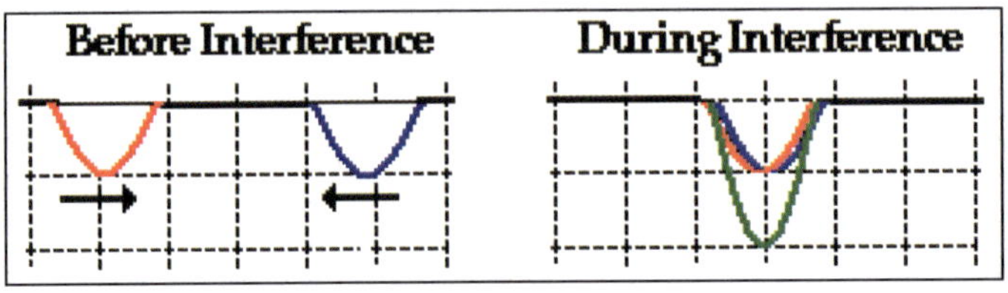

In this case, a sine pulse with a maximum displacement of -1 unit (negative means a downward displacement) interferes with a sine pulse with a maximum displacement of -1 unit. These two pulses are drawn in red and blue. The resulting shape of the medium is a sine pulse with a maximum displacement of -2 units.

Destructive Interference

Destructive interference is a type of interference that occurs at any location along the medium where the two interfering waves have a displacement in the opposite direction. For instance, when a sine pulse with a maximum displacement of +1 unit meets a sine pulse with a maximum displacement of -1 unit, destructive interference occurs. This is depicted in the diagram below.

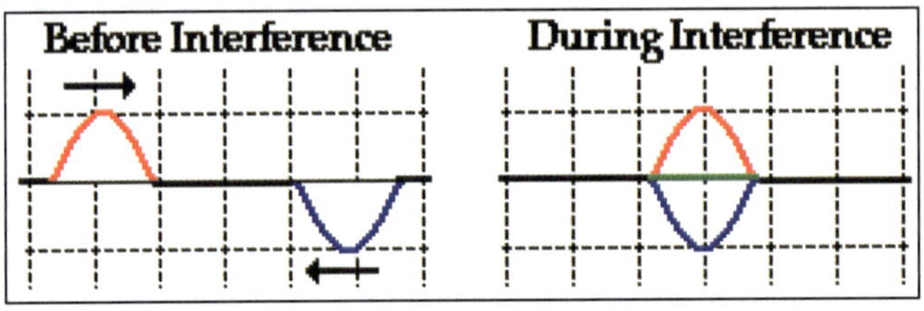

In the diagram above, the interfering pulses have the same maximum displacement but in opposite directions. The result is that the two pulses completely destroy each other when they are completely overlapped. At the instant of complete overlap, there is no resulting displacement of the particles of the medium. This “destruction” is not a permanent condition. In fact, to say that the two waves destroy each other can be partially misleading. When it is said that the two pulses destroy each other, what is meant is that when overlapped, the effect of one of the pulses on the displacement of a given particle of the medium is destroyed or canceled by the effect of the other pulse. Waves transport energy through a medium by means of each individual particle pulling upon its nearest neighbor. When two pulses with opposite displacements (i.e., one pulse displaced up and the other down) meet at a given location, the upward pull of one pulse is balanced (canceled or destroyed) by the downward pull of the other pulse. Once the two pulses pass through each other, there is still an upward displaced pulse and a downward displaced pulse heading in the same direction that they were heading before the interference. Destructive interference leads to only a momentary condition in which the medium’s displacement is less than the displacement of the largest-amplitude wave.

The two interfering waves do not need to have equal amplitudes in opposite directions for destructive interference to occur. For example, a pulse with a maximum displacement of +1 unit could meet a pulse with a maximum displacement of -2 units. The resulting displacement of the medium during complete overlap is -1 unit.

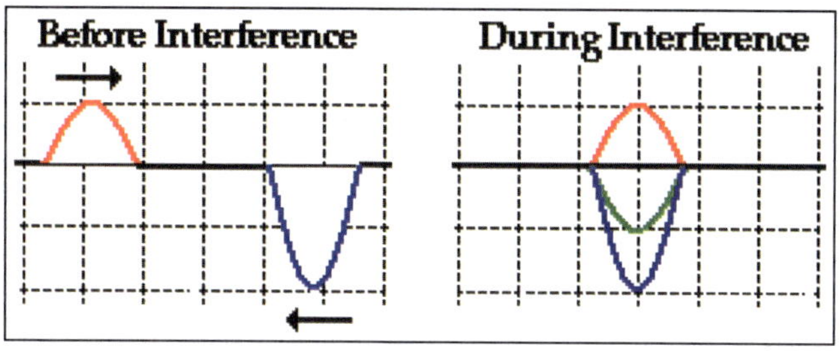

This is still destructive interference since the two interfering pulses have opposite displacements. In this case, the destructive nature of the interference does not lead to complete cancellation.

Interestingly, the meeting of two waves along a medium does not alter the individual waves or even deviate them from their path. This only becomes an astounding behavior when it is compared to what happens when two billiard balls meet or two football players meet. Billiard balls might crash and bounce off each other and football players might crash and come to a stop. Yet two waves will meet, produce a net resulting shape of the medium, and then continue on doing what they were doing before the interference.

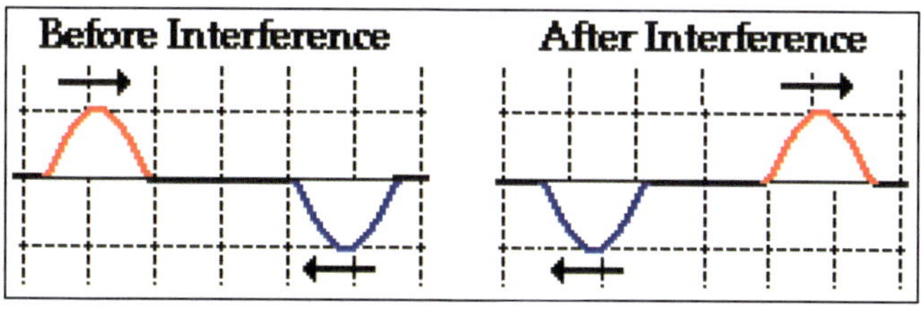

The Principle of Superposition

The task of determining the shape of the resultant demands that the principle of superposition is applied. The principle of superposition is sometimes stated as follows:

> When two waves interfere, the resulting displacement of the medium at any location is the algebraic sum of the displacements of the individual waves at that same location.

In the cases above, the summing the individual displacements for locations of complete overlap was made out to be an easy task - as easy as simple arithmetic:

Displacement of Pulse 1	Displacement of Pulse 2	=	Resulting Displacement
+1	+1	=	+2
-1	-1	=	-2

+1	-1	=	0
+1	-2	=	-1

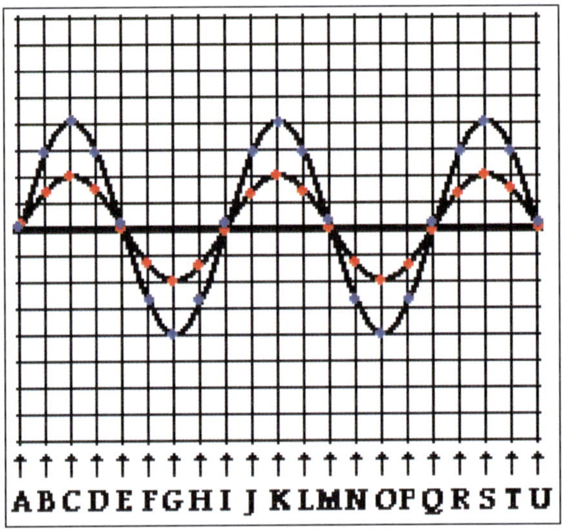

In actuality, the task of determining the complete shape of the entire medium during interference demands that the principle of superposition be applied for every point (or nearly every point) along the medium. As an example of the complexity of this task, consider the two interfering waves at the right. A snapshot of the shape of each individual wave at a particular instant in time is shown. To determine the precise shape of the medium at this given instant in time, the principle of superposition must be applied to several locations along the medium. A short cut involves measuring the displacement from equilibrium at a few strategic locations. Thus, approximately 20 locations have been picked and labeled as A, B, C, D, etc. The actual displacement of each individual wave can be counted by measuring from the equilibrium position up to the particular wave. At position A, there is no displacement for either individual wave; thus, the resulting displacement of the medium at position will be 0 units. At position B, the smaller wave has a displacement of approximately 1.4 units (indicated by the red dot); the larger wave has a displacement of approximately 2 units (indicated by the blue dot). Thus, the resulting displacement of the medium will be approximately 3.4 units. At position C, the smaller wave has a displacement of approximately 2 units; the larger wave has a displacement of approximately 4 units; thus, the resulting displacement of the medium will be approximately 6 units. At position D, the smaller wave has a displacement of approximately 1.4 units; the larger wave has a displacement of approximately 2 units; thus, the resulting displacement of the medium will be approximately 3.4 units. This process can be repeated for every position. When finished, a dot (done in green below) can be marked on the *graph* to note the displacement of the medium at each given location. The actual shape of the medium can then be sketched by estimating the position between the various marked points and sketching the wave. This is shown as the green line in the diagram below.

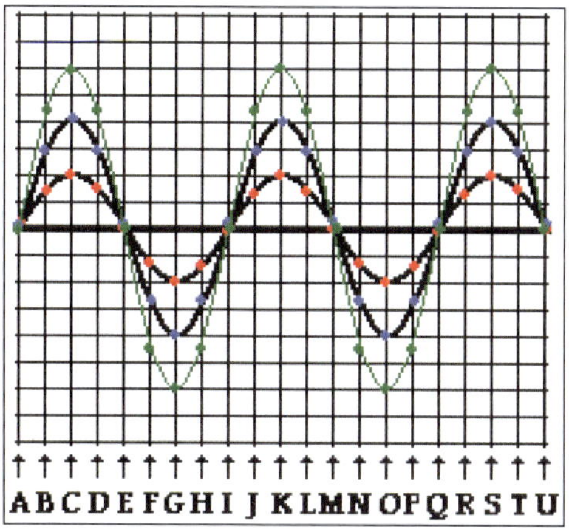

Interference of Light Waves

An important characteristic of light waves is their ability, under certain circumstances, to interfere with one another. Most people observe some type of optical interference every day, but do not realize what is occurring to produce this phenomenon. One of the best examples of interference is demonstrated by the light reflected from a film of oil floating on water. Another example is the soap bubble illustrated in figure that reflects a variety of beautiful colors when illuminated by natural or artificial light sources.

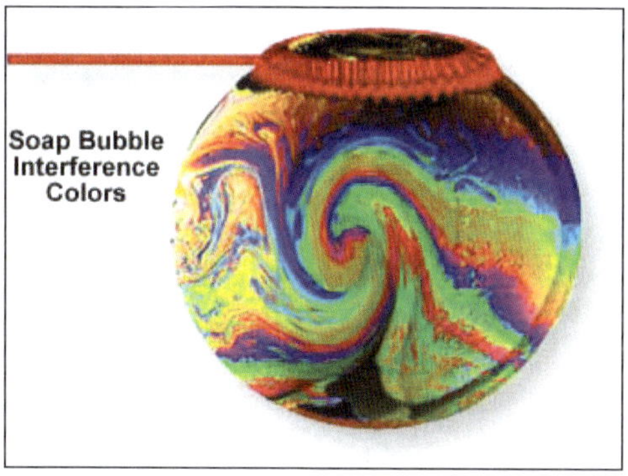

This dynamic interplay of colors derives from simultaneous reflection of light from both the inside and outside surfaces of the bubble. The two surfaces are very close together (they are only a few microns thick) and light reflected from the inner surface interferes both constructively and destructively with light reflected from the outer surface. This is because light reflected from the inner surface of the bubble must travel further than light reflected from the outer surface. When the waves reflected from the inner and outer surface combine they will interfere with each other, removing or reinforcing some parts of white light by destructive or constructive interference. This results in color. If the extra distance traveled by the inner light waves is exactly the wavelength of the outer light waves, then they will recombine constructively and bright colors of those wavelengths will be produced. In places where the waves are out of step, destructive interference will occur, canceling the reflected light (and the color).

The following is an explanation of how light waves interfere with each other. Consider a pair of light waves from the same source that are traveling, for example, in direction D. This is the propagation direction (as illustrated in figure) and if the vibrations (which are perpendicular to the propagation direction as represented by C in figure) are parallel to each other and are also parallel with respect to the direction of vibration, then the light waves may interfere with each other. If the vibrations are not in the same plane and are vibrating at 90 degrees to each other, then they cannot interfere with one another.

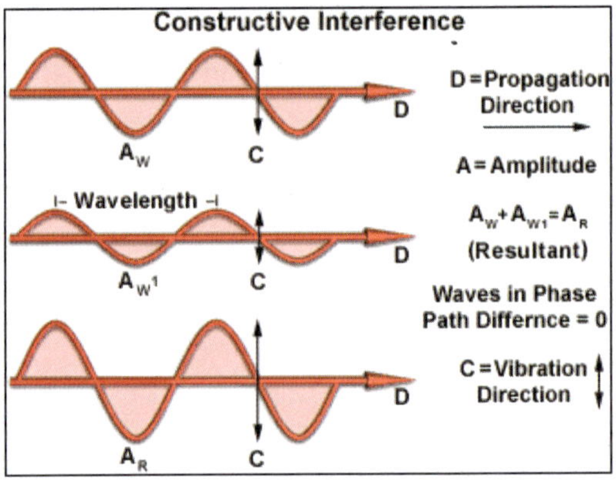

Assuming all of the criteria listed above are met, then the waves can interfere either constructively or destructively with each other. If the crests of one of the waves coincide with the crests of the other, the amplitudes are additive. If the amplitudes of both waves are equal, the resultant amplitude would be doubled. Bear in mind that light intensity varies directly as the square of the amplitude. Thus, if the amplitude is doubled, intensity is quadrupled. Such additive interference is called constructive interference (illustrated in figure).

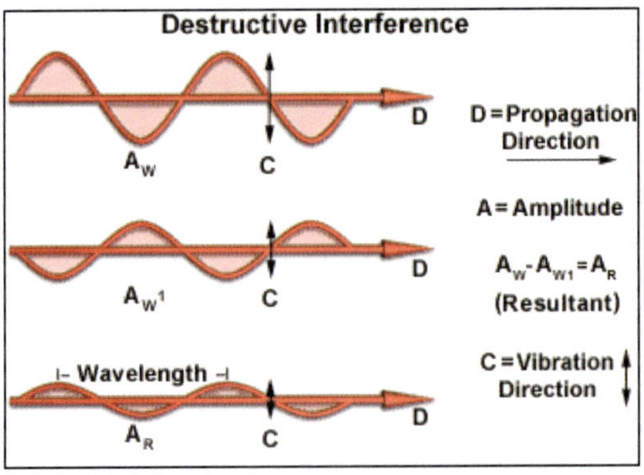

If the crests of one wave coincide with the troughs of the other wave, the resultant amplitude is decreased or may even be completely canceled, as illustrated in figure.

This is called destructive interference. The result is a drop in intensity, or in the case of total cancellation, blackness.

Thomas Young was a early 19th century physicist who demonstrated interference showing that light is a wave phenomenon and who also postulated that different colors of light were made from waves with different lengths. This was contrary to common opinion at the time, which was widely biased toward the theory that light is a stream of particles. In 1801, Young conducted an experiment providing important evidence that visible light has wave-like properties. This classic experiment, often termed “the Double-Slit experiment”, originally used sunlight that had first been diffracted through a single slit as a light source, but we will describe the experiment using coherent red laser light.

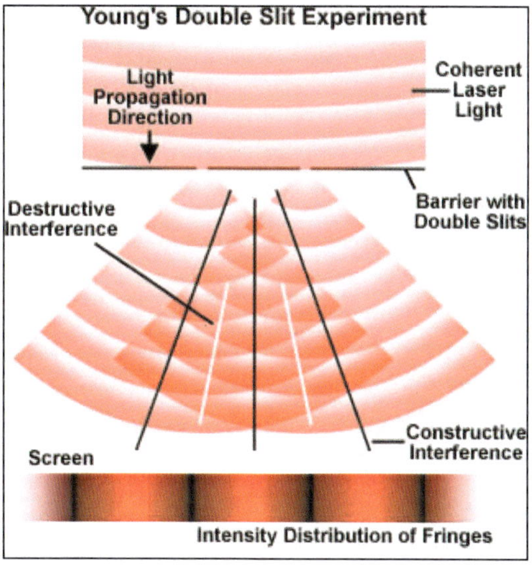

The basic setup of the double-slit experiment is illustrated in figure. Coherent laser light is allowed to illuminate a barrier containing two pinhole apertures that allow only some of the light to pass through. A screen is placed in the region behind the slits, and a pattern of bright red and dark interference bands becomes visible on the screen. The key to this experiment is the mutual coherence between the light diffracted from the two slits at the barrier. Young achieved this coherence through the diffraction of sunlight from the first slit.

As laser light is diffracted through the two barrier slits, each diffracted wave meets the other in a series of steps, as illustrated in figure. Sometimes the waves meet in step (or in phase; constructive interference), sometimes they meet out of step (or out of phase; destructive interference), and sometimes they meet partially in step. When the waves meet in step, they add together due to constructive interference and a bright area is displayed on the screen. In areas where the waves meet totally out of step, they will subtract from each other due to destructive interference and a dark area will appear in that portion of the screen. The resulting patterns on the screen, a product of interference between the two diffracted beams of laser light, are often referred to as interference fringes.

Other types of experiments have been devised to demonstrate the wave-like nature of light and interference effects. Most notable are the single mirror experiment of Lloyd and the double mirror and bi-prism experiments devised by Augustin Fresnel.

Sir Isaac Newton, the famous 17th century mathematician and physicist, was one of the first scientists to study interference phenomena. In his famous Newton's Rings experiment, he placed a convex lens of large curvature radius on a flat glass plate and applied pressure to hold the lens and glass plate together. When he viewed the plates through reflected sunlight, he observed a series of concentric light and dark highly colored bands of light similar to those illustrated in figure. Newton recognized that the rings indicated the presence of some degree of periodicity and used this observation to suggest a wave theory of light. Despite this, Newton regarded light as a stream of particles.

The rings occur because of a thin layer of air that exists between the curved convex and flat glass surfaces. Light reflected from the top and bottom surfaces of the glass is superimposed (combined) and produces interference patterns that appear as the colored rings. This principle is often used by lens manufacturers to test the uniformity of large polished surfaces.

Interference intensity distribution fringes (such as those observed in Young's double slit experiment) vary in intensity when they are presented on a uniform background. The visibility (V) of the intensity was defined by Albert Michelson, an early 20th century physicist, as the difference between the maximum and minimum intensity of a fringe divided by their sum:

$$V = I_{(max)} - I_{(min)}/I_{(max)} + I_{(min)}$$

Where $I_{(max)}$ is the maximum intensity and $I_{(min)}$ is the minimum intensity. From the equation, idealized fringe intensity always lies between zero and one, however in practice fringe visibility is dependent upon the geometrical design of the experiment and the spectral range used. This is responsible for the myriad of interference patterns observed in naturally occurring events.

Interference colors arising from stressed regions in materials can be easily observed in polarized light. The ruler in figure is made of plastic and is being observed through crossed polarizers. Under normal light, the ruler appears translucent with its graduations plainly visible. However, when viewed under polarized light, the ruler exhibits stress patterns that appear more profound in areas that are more highly deformed. This is due to a high degree of alignment of the long-chain polymer molecules that comprise the ruler. Note that the greatest degree of birefringence occurs near the hole on the left side of the ruler.

Birefringence Under Polarized Light

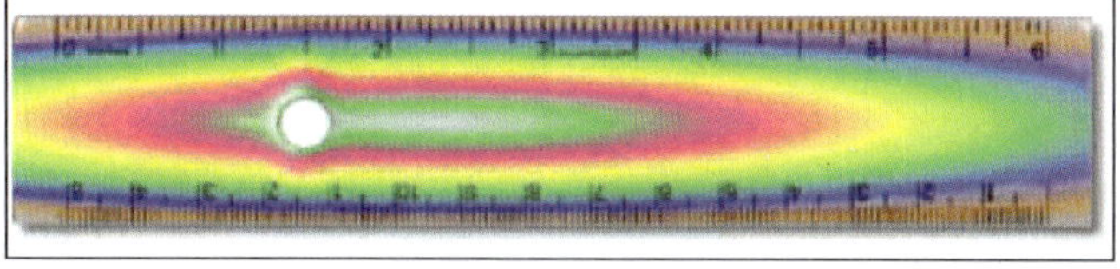

Other uses of interference are measurements made over long distances with lasers. In this case, the lasers can be used to measure very small distances over a range of many miles. This is accomplished by splitting the laser beam and reflecting it back from different surfaces. Analysis of the resulting interference fringes (upon recombining the separate laser beams) will yield a remarkably accurate calculation of the distance between the two objects.

Holograms also depend upon interference to produce their three-dimensional-like images. In reflection holograms, both a reference and object-illuminating beam are reflected onto a thick film from opposite sides. These beams interfere to produce light and dark areas that correspond to an image that appears three-dimensional. Transmission holograms use both the reference and object-illuminating beams on the same side of the film to produce a similar type of effect.

Interference also occurs with sound waves and waves induced in a standing pool of water. A very concise and easy interference experiment can be performed at home using a sink full of water and two marbles. First, let the water become very still, then simultaneously drop the marbles into the water (about 10-14 inches apart) from a height of about a foot. Just as with light waves, the two marbles will induce a series of waves in the water emanating in all directions. Waves formed in the area between where the marbles entered the water will eventually collide. Where they collide in step, they will constructively add together to make a bigger wave and where they collide out of step they will destructively cancel each other out.

Young's Double-slit Experiment

Young's double-slit experiment uses two coherent sources of light placed at a small distance apart, usually, only a few orders of magnitude greater than the wavelength of light is used. Young's double-slit experiment helped in understanding the wave theory of light which is explained with the help of a diagram. A screen or photodetector is placed at a large distance 'D' away from the slits as shown.

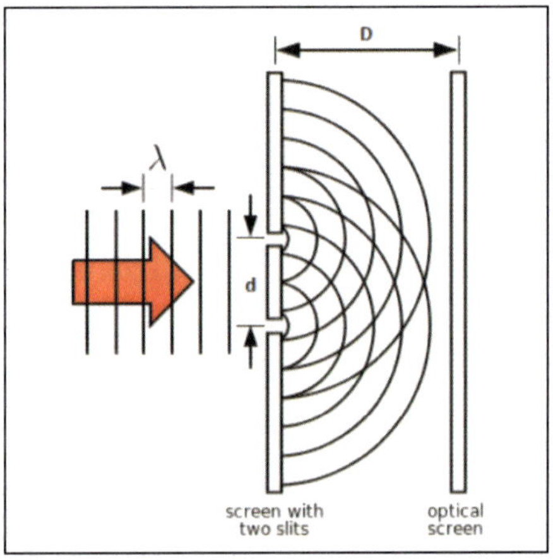

Derivation of Young's Double Slit Experiment

The original Young's double-slit experiment used diffracted light from a single source passed into two more slits to be used as coherent sources. Lasers are commonly used as coherent sources in the modern-day experiments.

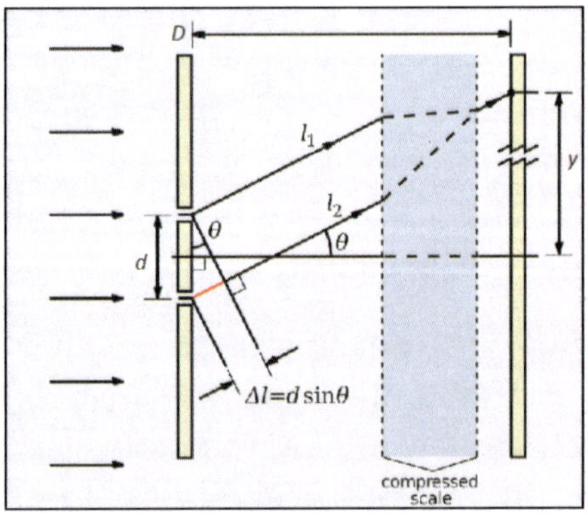

Young's Double Slit Experiment

Each source can be considered as a source of coherent light waves. At any point on the screen at a distance 'y' from the centre, the waves travel distances l1 and l2 to create a path difference of Δl at the point in question. The point approximately subtends an angle of θ at the sources (since the distance D is large there is only a very small difference of the angles subtended at sources).

Derivation of Young's Double Slit Experiment

Consider a monochromatic light source 's' kept at a considerable distance from two slits s_1 and s_2. S is equidistant from s_1 and s_2. s_1 and s_2 behave as two coherent sources, as both bring derived from S.

The light passes through these slits and falls on a screen which is at a distance 'D' from the position of slits s_1 and s_2. 'd' be the separation between two slits.

If s_1 is open and s_2 is closed, the screen opposite to s_1 is closed, only the screen opposite to s_2 is illuminated. The interference patterns appear only when both slits s_1 and s_2 are open.

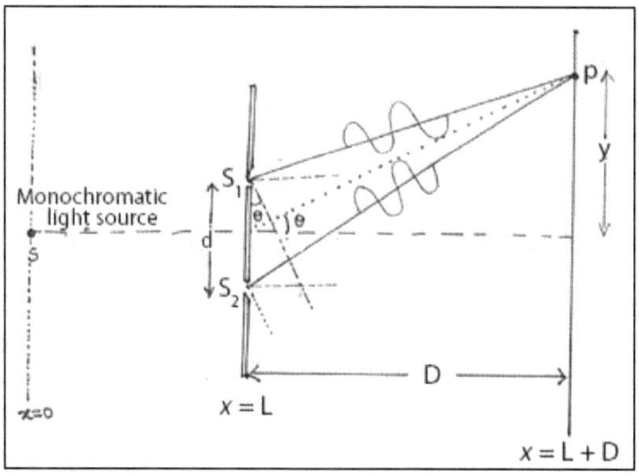

When the slit separation (d) and the screen distance (D) are kept unchanged, to reach *P* the light waves from s_1 and s_2 must travel different distances. It implies that there is a path difference in Young's double slit experiment between the two light waves from s_1 and s_2.

Approximations in Young's Double Slit Experiment

- Approximation 1:D > > d: Since D > > d, the two light rays are assumed to be parallel, then the path difference,
- Approximation 2: d/λ >> 1: Often, d is a fraction of a millimetre and λ is a fraction of a micrometre for visible light.

Under these conditions θ is small, thus we can use the approximation sin θ approx tan θ = γ/D.

$\therefore$ Path difference, $\Delta z = \gamma / D$.

This is the path difference between two waves meeting at a point on the screen. Due to this path difference in Young's double slit experiment, some points on the screen are bright and some points are dark.

Position of Fringes in Young's Double Slit Experiment

Position of Bright Fringes

For maximum intensity at P,

$\Delta z = n\lambda$ (n = 0, ±1, ±2, ...)

Or d sin θ = nλ (n = 0, ±1, ±2, ...)

The bright fringe for n = o is known as the central fringe. Higher order fringes are situated symmetrically about the central fringe. The position of nth bright fringe is given by,

y (bright) = (nλ\d)D (n = o, ±1, ±2, ...)

Position of Dark Fringes

For minimum intensity at P,

$$\Delta z = (2n-1)\frac{n\lambda}{2}(n = \pm1, \pm2, ...)$$

$$d\sin\theta = (2n-1)\frac{n\lambda}{2}$$

The first minima are adjacent to the central maximum on either side. We will obtain the position of dark fringe as,

$$y_{dark} = \frac{(2n-1)\lambda D}{2d}(n = \pm1, \pm2, ...)$$

Fringe Width

Distance between two adjacent bright (or dark) fringes is called the fringe width.

$$\beta z \frac{n\lambda D}{d} - \frac{(n-1)\lambda D}{d} = \frac{\lambda D}{d}$$

$$\Rightarrow \beta \propto \lambda$$

If the apparatus of Young's double slit experiment is immersed in a liquid of refractive index (u), then wavelength of light and hence fringe width decreases 'u' times.

$$\beta^1 = \frac{\beta}{\mu}$$

If a white light is used in place of monochromatic light, then coloured fringes are obtained on the screen with red fringes larger in size that of violet.

Angular Width of Fringes

Let the angular position of nth bright fringe is θ_n and because of its small value,

$$\tan\theta_n \approx \theta_n$$

$$\tan\theta_n = \frac{\gamma n}{D} \approx \theta_n = \frac{\gamma n}{D}$$

$$\Rightarrow \theta_n = \frac{n\gamma D/d}{D} = \frac{n\lambda}{d}$$

Similarly, the angular position of $(n+1)^{th}$ bright fringe is θ_{n+1}, then,

$$\theta_{n+1} = \frac{\gamma_{n+1}}{D} = \frac{(n+1)\lambda D/d}{D} = \frac{(n+1)\lambda}{d}$$

∴ Angular width of a fringe in Youngs double slit experiment is given by,

$$\theta = \theta_{n+1} - \theta_n = \frac{(n+1)\lambda}{d} - \frac{n\lambda}{d} = \frac{\lambda}{d}\theta = \frac{\lambda}{d}$$

We know that $\beta = \frac{\lambda D}{d}$

$$\Rightarrow \theta = \frac{\lambda}{d} = \frac{\beta}{D}$$

Angular width is independent of 'n' i.e angular width of all fringes are same.

Maximum Order of Interference Fringes

The position of nth order maxima on the screen is $\gamma = \frac{n\lambda D}{d}; n = 0, \pm 1, \pm 2, ...$

But 'n' values cannot take infinitely large values as it would violate 2nd approximation.

i.e θ is small (or) y < < D

$$\Rightarrow \frac{\gamma}{D} = \frac{n\lambda}{d} << 1$$

Hence, the above formula for interference maxima is applicable when $n << \frac{d}{\lambda}$.

When 'n' value becomes comparable to $\frac{d}{\lambda}$, path difference can no longer be given by $\frac{d\gamma}{D}$,Hence for maxima, path difference = nλ.

$$\Rightarrow d\sin\theta = n\lambda$$

$$\Rightarrow n = \frac{d\sin\theta}{\lambda} n_{max} = \left[\frac{d}{\lambda}\right]$$

The above represents box function or greatest integer function.

Similarly, the highest order of interference minima,

$$n_{min} = \left[\frac{d}{\lambda} + \frac{1}{2}\right]$$

Shape of Interference Fringes in YDSE

From the given YDSE diagram, the path difference from the two slits is given by,

$$s_2p - s_1p \approx d \sin\theta \text{ (constant)}$$

The above equation represents a hyperbola with its two foci as s_1 and s_2.

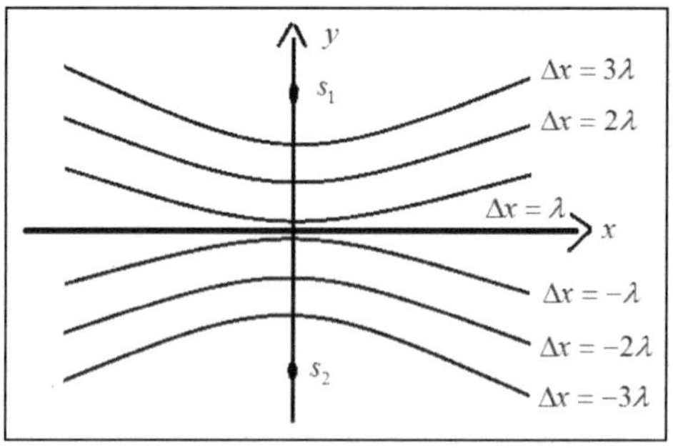

The interference pattern we get on the screen is a section of a hyperbola when we revolve hyperbola about the axis s_1s_2.

If the fringe will represent 1st minima, the fringe will represent 1st maxima, it represents central maxima.

If the screen is yz plane, fringes are hyperbolic with a straight central section.

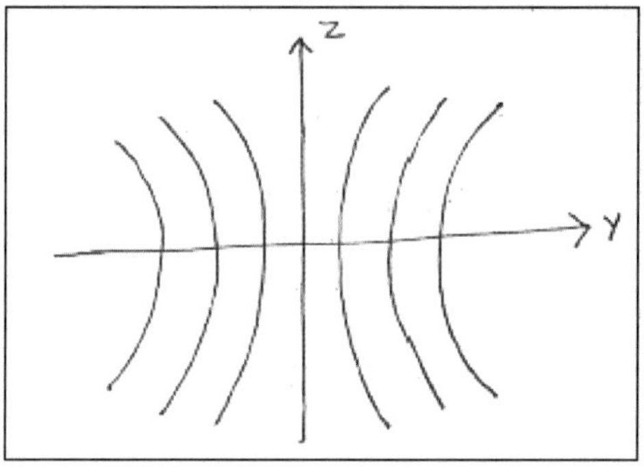

If the screen is xy plane, the fringes are hyperbolic with a straight central section.

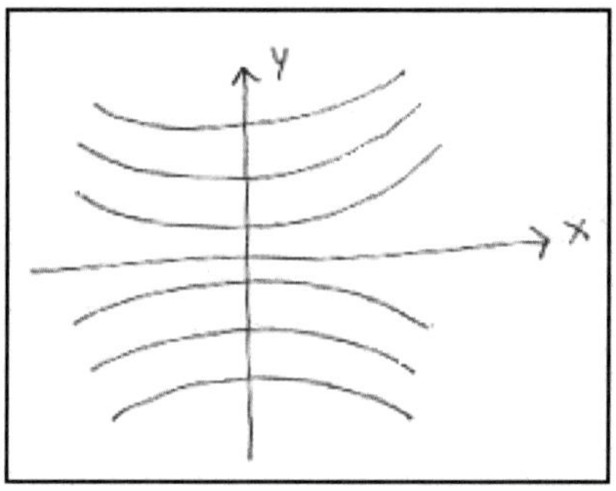

Intensity of Fringes in Young's Double Slit Experiment

For two coherent sources s_1 and s_2, the resultant intensity at point *p* is given by,

$$I = I_1 + I_2 + 2\sqrt{(I_1 . I_2)} \cos \varphi$$

Putting $I_1 = I_2 = I_0$ (Since, d<<<D)

$$I = I_0 + I_0 + 2\sqrt{(I_0 . I_0)} \cos \varphi$$

$$I = 2I_0 + 2(I_0) \cos \varphi$$

$$I = 2I_0 (H \cos s \varphi)$$

$$I = 4I_0 \cos^2\left(\frac{\phi}{2}\right).$$

For Maximum Intensity

$$\cos\frac{\phi}{2} = \pm 1 \frac{\phi}{2} = n\pi, n = 0, \pm 1, \pm 2, \ldots\ldots$$

$$\text{or } \varphi = 2n\pi$$

Phase difference $\varphi = 2n\pi$

then, path difference $\Delta x = \frac{\lambda}{2\pi}(2n\pi) = n\lambda$

The intensity of bright points are maximum and given by,

$$I_{max} = 4I_0$$

For Minimum Intensity

$$\cos\frac{\phi}{2} = 0 \frac{\phi}{2} = \left(n - \frac{1}{2}\right)\pi \text{ where } (n = \pm 1, \pm 2, \pm 3, \ldots\ldots)$$

$$\varphi = (2n-1)\pi$$

Phase difference $\varphi = (2n - 1)\pi$

$$\frac{2\pi}{\lambda}\Delta x = (2n-1)\pi \quad \Delta x = (2n-1)\frac{\lambda}{2}$$

Thus, intensity of minima is given by,

$I_{min} = 0$

If $I_1 \neq I_2, I_{min} \neq 0.$

Special Cases

Rays Not Parallel to Principal Axis

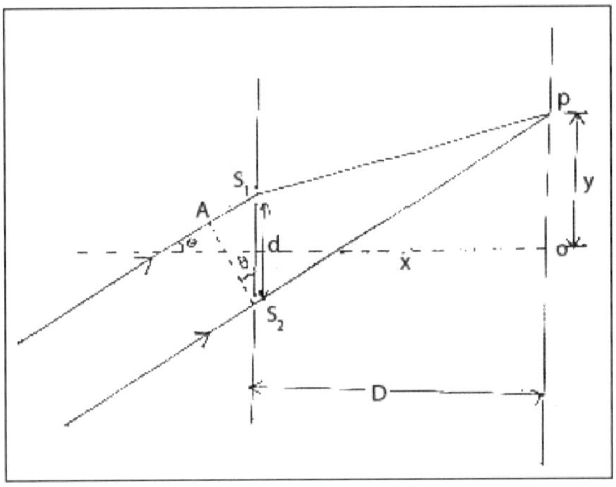

From the above diagram,

Path difference $\Delta x = \left(AS_1 + S_1P\right) - S_2P \Delta x = AS_1 - \left(S_2P - S_1P\right)\Delta x = d\sin\theta - \dfrac{4d}{D}$

For maxima $\Delta x = n\lambda$

For minima $\Delta x = \left(2n-1\right)\dfrac{\lambda}{2}$

Using this we can calculate different positions of maxima and minima.

Source Placed beyond the Central Line

If the source is placed a little above or below this centre line, the wave interaction with S_1 and S_2 has a path difference at a point P on the screen,

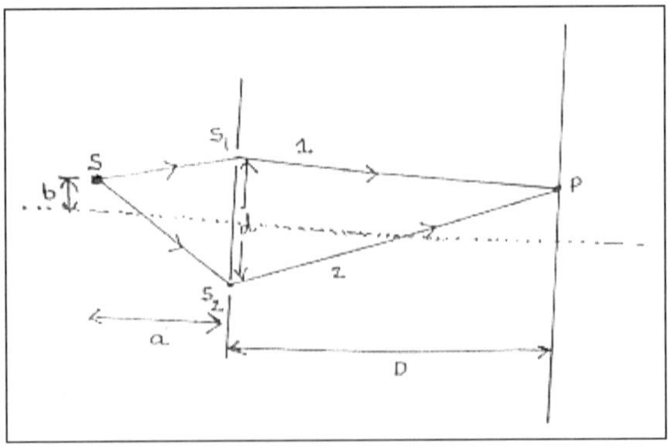

Δ x= (distance of ray 2) – (distance of ray 1)

$$= \left(S\ S_2 + S_2P\right) - \left(S\ S_1 + S_1P\right)$$
$$= \left(S\ S_2 + S\ S_1\right) + \left(S_2P - S_1P\right)$$

$= bd/a + yd/D \rightarrow (*)$

We know $\Delta x = n\lambda$ for maximum.

$\Delta x = (2n - 1)\lambda/2$ for minimum.

By knowing the value of Δx from (*) we can calculate different positions of maxima and minima.

Displacement of Fringes in YDSE

When a thin transparent plate of thickness 't' is introduced in front of one of the slits in Young's double slit experiment, the fringe pattern shifts toward the side where the plate is present.

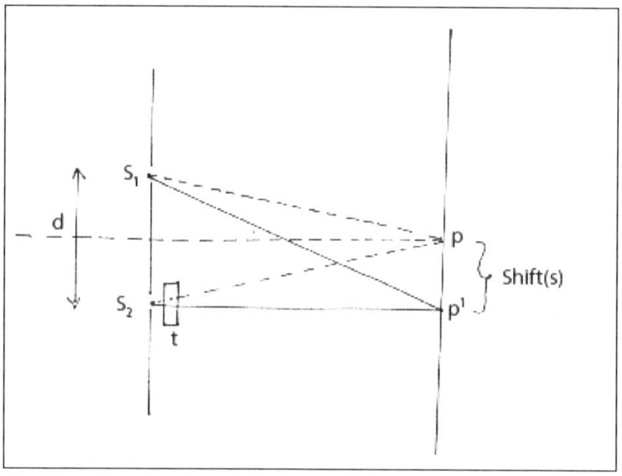

The dotted lines denote the path of the light before introducing the transparent plate. The solid lines denote the path of the light after introducing a transparent plate.

Path difference before introducing the plate $\Delta x = S_1P - S_2P$.

Path difference after introducing the plate $\Delta x_{new} = S_1P^1 - S_2P^1$.

The path length $S_2P^1 = \left(S_2P^1 - t\right)_{air} + t_{plate} = \left(S_2P^1 - t\right)_{air} + \left(\mu t\right)_{plate}$.

Where $'\mu t'$ is optical path,

$$= S_2P^1{}_{air} + (\mu - 1)t$$

Then we get,

$$(\Delta x)_{new} = S_1P^1{}_{air} - \left(S_2P^1{}_{air} + (\mu - 1)t\right) = \left(S_1P^1 - S_2P^1\right)_{air} - (\mu - 1)t$$

$$(\Delta x)_{new} = d\sin\theta - (\mu - 1)t \quad (\Delta x)_{new} = \frac{\gamma d}{D} - (\mu - 1)t$$

Then,

$$y = \frac{\Delta x D}{d} + \frac{D}{d}\left[(\mu - 1)t\right]$$
$$\downarrow \qquad \qquad \downarrow$$
$$(1) \qquad \qquad (2)$$

Term (1) defines the position of a bright or dark fringe, the term (2) defines the shift occurred in the particular fringe due to the introduction of a transparent plate.

Constructive and Destructive Interference

For constructive interference, the path difference must be an integral multiple of the wavelength.

Thus for a bright fringe to be at 'y',

nλ = y dD

Or, ynth = nλ Dd

Where n = ± 0, 1, 2, 3.....

The 0th fringe represents the central bright fringe.

Similarly, the expression for a dark fringe in Young's double slit experiment can be found by setting the path difference as:

$$\Delta \mathrm{I} = (\mathrm{n} + 12)\lambda$$

This simplifies to $yn = (n + 12)\lambda Dd$.

Note that these expressions require that θ be very small. Hence yD needs to be very small. This implies D should be very large and y should be small. This, in turn, requires that the formula works best for fringes close to the central maxima. In general, for best results, dD must be kept as small as possible for a good interference pattern.

The Young's double slit experiment was a watershed moment in scientific history because it firmly established that light indeed behaved as a wave.

The Double Slit Experiment was later conducted using electrons, and to everyone's surprise, the pattern generated was similar as expected with light. This would forever change our understanding of matter and particles, forcing us to accept that matter like light also behaves like a wave.

Fresnel Biprism

A Fresnel Biprism is a thin double prism placed base to base and have very small refracting angle (0.5°). This is equivalent to a single prism with one of its angle nearly 179° and other two of 0.5° each.

The interference is observed by the division of wave front. Monochromatic light through a narrow slit S falls on biprism , which divides it into two components. One of these component is refracted from upper portion of biprism and appears to come from S_1 where the other one refracted through lower portion *and* appears to come from S_2. Thus S_1 and S_2 act as two virtual coherent sources formed from the original source. Light waves arising from S_1 and S_2 interfere in the shaded region and interference fringes are formed which can be observed on the screen.

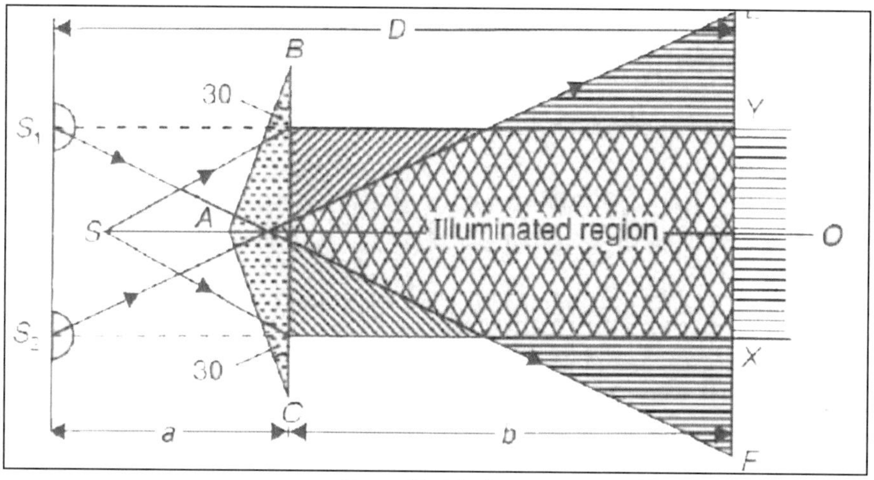

Fresnel's Biprism.

Applications of Fresnel's Biprism

Fesnel biprism can be used to determine the wavelength of a light source (monochromatic), thickness of a thin transparent sheet/thin film, refractive index of medium etc.

Determination of Wave Length of light

As expression for fringe width is $\beta = \frac{D\lambda}{d}$.

Biprism can be used to determine the wavelength of given monochromatic light using the expression:

$$\lambda = \frac{d\beta}{D}$$

Experimental Arrangement: Light from monochromatic source is made to fall on a thin slit mounted vertically on a rigid optical bench fitted with a scale. The biprism and the screen (in this case an eye piece) are also mounted vertically. The eye piece can be moved in the plane perpendicular to the axis of bench using a micrometer based translation stage.

(i) Measurement of fringe width: To get β, fringes are first observed in the field of view of the microscope. The vertical wire of the eyepiece is made to coincide with one of the fringes and screw of micrometer is moved sideways and number of fringes is counted.

β =Distance moved/number of fringes passed

(ii) Measurement of D: This distance between source and eyepiece is directly measured on the optical bench scale.

(iii) Determination of d: To determine the separation between the two virtual sources (d), a convex lens of short focal length is introduced between the biprism and the eye piece, keeping the distance between the slit and eyepiece to be more than four times the focal length of lens. The lens is moved along the length of bench to a position where two images of slits are seen in the plane of cross wires of eye piece. The distance between these two images of slit is measured by setting the vertical cross wire successively on each of images and recording the two positions of cross wire using micrometer. Let this separation be d_1 . Now the lens is moved such that for another position of lens, again two images of slit are seen on eye piece. Let d_2 be the separation between these two images.

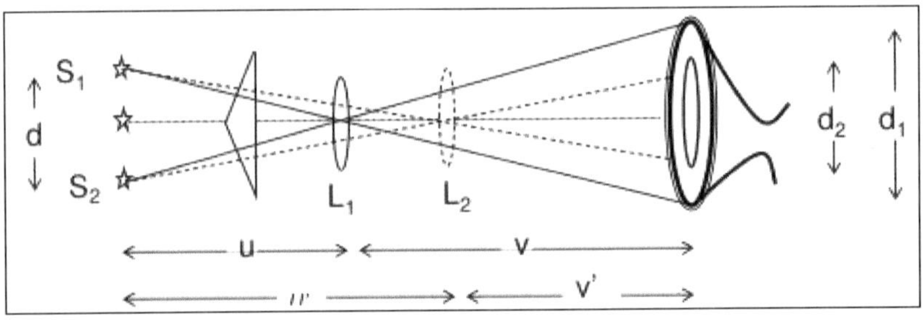

$$\frac{d_1}{d} = \frac{v}{u}$$

$$\frac{d_2}{d} = \frac{v'}{u'}$$

Determination of Thickness of a Thin Film

To determine the thickness of transparent thin sheet (mica), the monochromatic source is replaced by white light source.

The position of this central white fringe is recorded by focusing the cross wire of eye piece on it and taking this reading of micrometer scale. Now mica sheet is introduced in the path of one wave. (such that it blocks the one half of biprism). By doing it the one wave traverse an extra optical path and the path difference between the two waves is not same and entire fringe pattern shifts. The central white fringe is now shifted to another position of cross wire. If '*S*' is the shift in position of white fringe and be the refractive index of mica sheet, thickness '*t*' of mica sheet is given by:

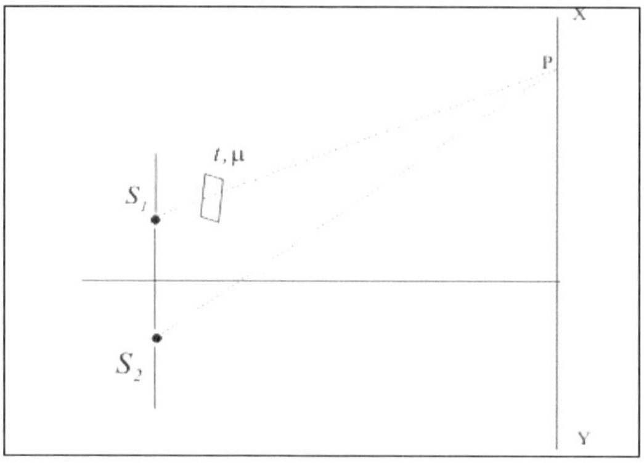

Thin film is introduced in path of one of source of light.

Time required by light to reach from S_1 to point P,

$$=\frac{S_1P-t}{c}+\frac{t}{v}$$

Where v = c/μ

$$T=\frac{S_1P+t(\mu-1)}{c}$$

Hence equivalent path that is covered by light in air is $S_1P+t(\mu-1)$,

Optical path difference at P,

$$=S_2P-[S_1P+\mu t-t]$$
$$=S_2P-S_1P-[\mu-]t$$
$$=\frac{xd}{D}-[\mu-]t$$

Therefore n^{th} fringe shift is given by,

$$\Delta x=\frac{D(\mu-1)t}{d}$$

As,

$$\beta=\frac{D}{d}\lambda$$
$$\Delta x=\frac{\beta(\mu-1)t}{\lambda}$$

Where λ is the wavelength of the wave; Δx is displacement of fringes and β is fringe width.

Displacement of Fringes

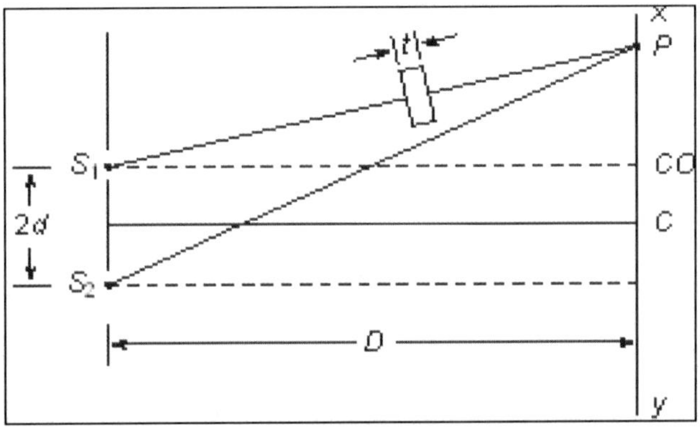

If a thin glass or mica strip or any other transparent plate of uniform thickness is introduced in the path of one of the two interfering beam from two coherent sources then central bright fringe will be displaced. This displacement from C to Coo will towards the side of lamina.

In order to calculate the displace we shall find the path difference between two beam from coherent sources S_1 and S_2 at any point 'P' on the screen. Let 't' be the thickness of sheet and 'μ' the refractive index of its materials.

Time T taken by the beam to reach from S_1 to P is given by,

$$T = \frac{S_1P}{C} + \frac{t}{V}$$

Where C is the velocity in air and V is velocity of light in the medium of the plate,

$$T = \frac{S_1P - t}{C} + \frac{\mu t}{C} \qquad \therefore \mu = \frac{C}{V}$$

$$T = \frac{S_1P - t + \mu t}{C} = \frac{S_1P + (\mu - 1)t}{C}$$

The effective path in air from S_1 to P is $S_1\ P + (\mu + 1)$t so the air path S_1 P has been increased,

By $(\mu + 1)$t as a result of the introduction of the plate.

The path difference between reaching P from S_1 to P

= path covered by beam S_2 to P – optical path from S_1 to P

$$= S_2P - \left[S_1P + (\mu - 1)t\right] = S_2P - S_1P - (\mu - 1)t$$

But we know that,

$$S_2P - S_1P = \frac{2yd}{D}$$

Path Difference $\Rightarrow \delta = \frac{2yd}{D} - (\mu - 1)t$

Condition for Maxima

If P is the centre of the nth bright fringe.

$$\frac{2y_n d}{D} - (\mu - 1)t = n\lambda \qquad n = 0, 1, 2, 3$$

$$\frac{2y_n d}{D} = n\lambda + (\mu - 1)t$$

$$y_n \frac{D}{2d}\left[n\lambda + (\mu - 1)t\right]$$

Shift of Central Fringe

Shift of central bright fringe where n=0 is given by,

$$y_0 = \frac{D}{2d}(\mu - 1)t$$

This expression is independent of n. This indicate that all the bright fringes are displaced through the same amount equal to similarly it can be shown that the displacement of any dark fringe is also,

$$= \frac{D}{2d}(\mu - 1)t$$

$$y_0 = \frac{D}{2d}(\mu - 1)t$$

The fringe width ω is given by,

$$\omega = y_{n+1} - y_n$$

$$\omega = \frac{D\lambda}{2d}.$$

Lloyd's Mirror

Lloyd's mirror is an optics experiment that was first described in 1834 by Humphrey Lloyd in the Transactions of the Royal Irish Academy. Its original goal was to provide further evidence for the wave nature of light, beyond those provided by Thomas Young and Augustin-Jean Fresnel. In the experiment, light from a monochromatic slit source reflects from a glass surface at a small angle and appears to come from a virtual source as a result. The reflected light interferes with the direct light from the source, forming interference fringes. It is the optical wave analogue to a sea interferometer.

Setup

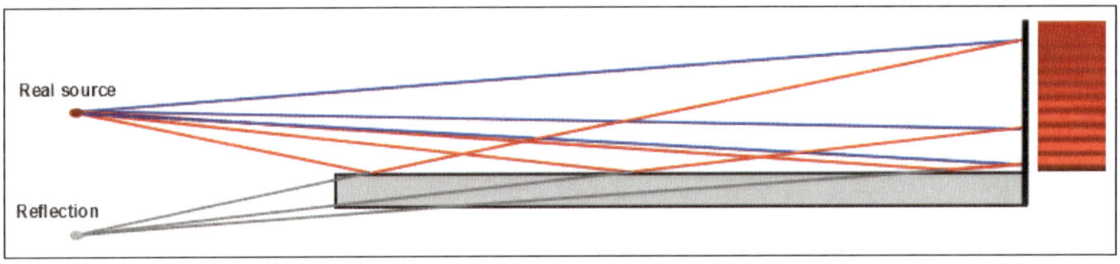

Figure: Lloyd's mirror.

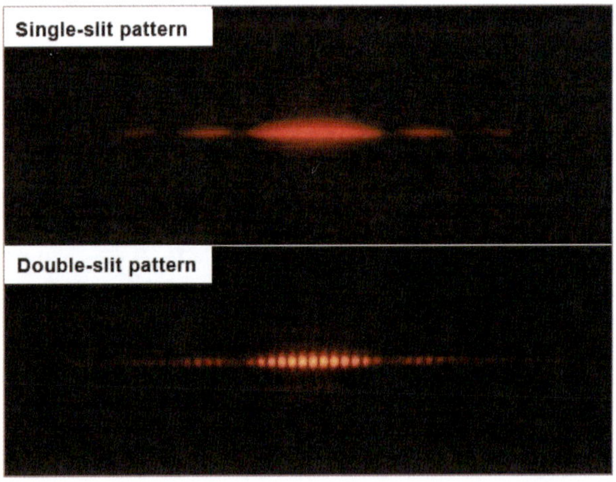

Young's two-slit experiment displays a single-slit diffraction pattern on top of the two-slit interference fringes.

Lloyd's Mirror is used to produce two-source interference patterns that have important differences from the interference patterns seen in Young's experiment.

In a modern implementation of Lloyd's mirror, a diverging laser beam strikes a front-surface mirror at a grazing angle, so that some of the light travels directly to the screen (blue lines in figure), and some of the light reflects off the mirror to the screen (red lines). The reflected light forms a virtual second source that interferes with the direct light.

In Young's experiment, the individual slits display a diffraction pattern on top of which is overlaid interference fringes from the two slits. In contrast, the Lloyd's mirror experiment does not use slits and displays two-source interference without the complications of an overlaid single-slit diffraction pattern.

In Young's experiment, the central fringe representing equal path length is bright because of constructive interference. In contrast, in Lloyd's mirror, the fringe nearest the mirror representing equal path length is dark rather than bright. This is because the light reflecting off the mirror undergoes a 180° phase shift, and so causes destructive interference when the path lengths are equal or when they differ by an integer number of wavelengths.

Applications

Interference Lithography

The most common application of Lloyd's mirror is in UV photolithography and nanopatterning. Lloyd's mirror has important advantages over double-slit interferometers. If one wishes to create a series of closely spaced interference fringes using a double-slit interferometer, the spacing d between the slits must be increased. Increasing the slit spacing, however, requires that the input beam be broadened to cover both slits. This results in a large loss of power. In contrast, increasing d in the Lloyd's mirror technique does not result in power loss, since the second "slit" is just the reflected virtual image of the source. Hence, Lloyd's mirror enables the generation of finely detailed interference patterns of sufficient brightness for applications such as photolithography.

Typical uses of Lloyd's mirror photolithography would include fabrication of diffraction gratings for surface encoders and patterning the surfaces of medical implants for improved biofunctionality.

Test Pattern Generation

High visibility $\cos^2$-modulated fringes of constant spatial frequency can be generated in a Lloyd's mirror arrangement using parallel collimated monochromatic light rather than a point or slit source. The uniform fringes generated by this arrangement can be used to measure the modulation transfer functions of optical detectors such as CCD arrays to characterize their performance as a function of spatial frequency, wavelength, intensity, and so forth.

Optical Measurement

The output of a Lloyd's mirror was analyzed with a CCD photodiode array to produce a compact, broad range, high accuracy Fourier transform wavemeter that could be used to analyze the spectral output of pulsed lasers.

Radio Astronomy

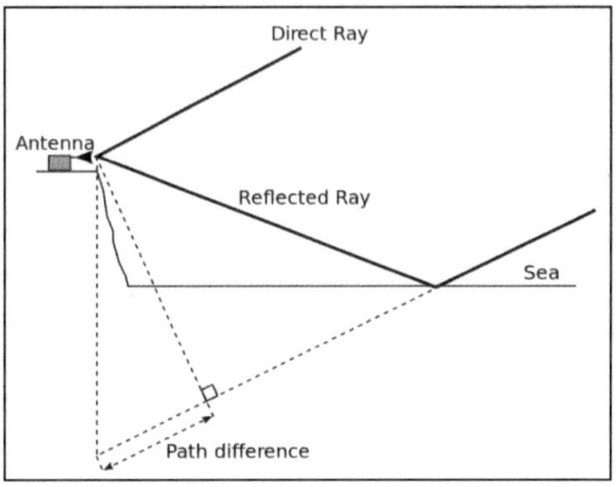

Determining the position of galactic radio sources using Lloyd's mirror.

In the late 1940s and early 1950s, CSIRO scientists used a technique based on Lloyd's mirror to make accurate measurements of the position of various galactic radio sources from coastal sites in New Zealand and Australia. As illustrated in figure, the technique was to observe the sources combining direct and reflected rays from high cliffs overlooking the sea. After correcting for atmospheric refraction, these observations allowed the paths of the sources above the horizon to be plotted and their celestial coordinates to be determined.

Underwater Acoustics

An acoustic source just below the water surface generates constructive and destructive interference between the direct path and reflected paths. This can have a major impact on sonar operations.

The Lloyd mirror effect has been implicated as having an important role in explaining why marine animals such as manatees and whales have been repeatedly hit by boats and ships. Interference due to Lloyd's mirror results in low frequency propeller sounds not being discernible near the surface, where most accidents occur. This is because at the surface, sound reflections are nearly 180 degrees out of phase with the incident waves. Combined with spreading and acoustic shadowing effects, the result is that the marine animal is unable to hear an approaching vessel before it has been run over or entrapped by the hydrodynamic forces of the vessel's passage.

Interference in Thin Films

The bright colors seen in an oil slick floating on water or in a sunlit soap bubble are caused by interference. The brightest colors are those that interfere constructively. This interference is between light reflected from different surfaces of a thin film; thus, the effect is known as thin-film interference.

Interference effects are most prominent when light interacts with something having a size similar to its wavelength. A thin film is one having a thickness t smaller than a few times the wavelength of light, λ. Since color is associated indirectly with λ and because all interference depends in some way on the ratio of λ to the size of the object involved, we should expect to see different colors for different thicknesses of a film, as in figure below.

These soap bubbles exhibit brilliant colors when exposed to sunlight.

Figure shows how light reflected from the top and bottom surfaces of a film can interfere. Incident light is only partially reflected from the top surface of the film (ray 1). The remainder enters the film and is itself partially reflected from the bottom surface. Part of the light reflected from the bottom surface can emerge from the top of the film (ray 2) and interfere with light reflected from the top (ray 1). The ray that enters the film travels a greater distance, so it may be in or out of phase with the ray reflected from the top. However, consider for a moment, again, the bubbles in figure. The bubbles are darkest where they are thinnest. Furthermore, if you observe a soap bubble carefully, you will note it gets dark at the point where it breaks. For very thin films, the difference in path lengths of rays 1 and 2 in figure is negligible, so why should they interfere destructively and not constructively? The answer is that a phase change can occur upon reflection.

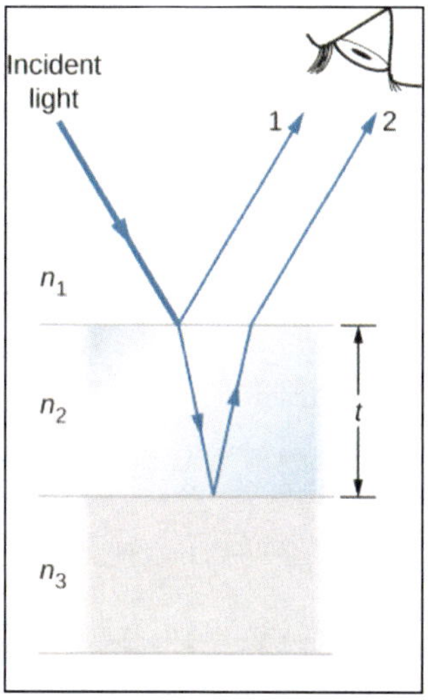

Light striking a thin film is partially reflected (ray 1) and partially refracted at the top surface. The refracted ray is partially reflected at the bottom surface and emerges as ray 2. These rays interfere in a way that depends on the thickness of the film and the indices of refraction of the various media.

Changes in Phase due to Reflection

Reflection of mechanical waves can involve a 180° phase change. For example, a traveling wave on a string is inverted (i.e., a 180° phase change) upon reflection at a boundary to which a heavier string is tied. However, if the second string is lighter (or more precisely, of a lower linear density), no inversion occurs. Light waves produce the same effect, but the deciding parameter for light is the index of refraction. Light waves undergo a 180° or π radians phase change upon reflection at an interface beyond which is a medium of higher index of refraction. No phase change takes place when reflecting from a medium of lower refractive index. Because of the periodic nature of waves, this phase change or inversion is equivalent to ± λ/2 in distance travelled, or path length. Both the path length and refractive indices are important factors in thin-film interference.

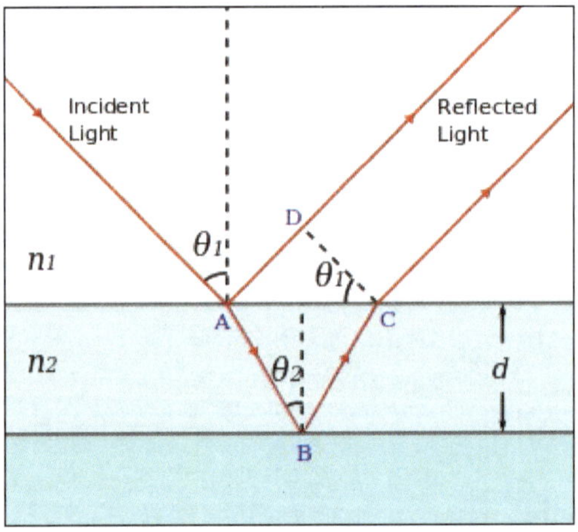

If the film in figure is a soap bubble (essentially water with air on both sides), then a phase shift of λ/2 occurs for ray 1 but not for ray 2. Thus, when the film is very thin and the path length difference between the two rays is negligible, they are exactly out of phase, and destructive interference occurs at all wavelengths. Thus, the soap bubble is dark here. The thickness of the film relative to the wavelength of light is the other crucial factor in thin-film interference. Ray 2 in figure travels a greater distance than ray 1. For light incident perpendicular to the surface, ray 2 travels a distance approximately 2t farther than ray 1. When this distance is an integral or half-integral multiple of the wavelength in the medium ($\lambda_n = \lambda / n$, where λ is the wavelength in vacuum and n is the index of refraction), constructive or destructive interference occurs, depending also on whether there is a phase change in either ray.

Calculating the Thickness of a Nonreflective Lens Coating

Sophisticated cameras use a series of several lenses. Light can reflect from the surfaces of these various lenses and degrade image clarity. To limit these reflections, lenses are coated with a thin layer of magnesium fluoride, which causes destructive thin-film interference. What is the thinnest this film can be, if its index of refraction is 1.38 and it is designed to limit the reflection of 550-nm light, normally the most intense visible wavelength? Assume the index of refraction of the glass is 1.52.

Strategy

Refer to figure given below and use $n_1 = 1.00$ for air, $n_2 = 1.38$, and $n_3 = 1.52$. Both ray 1 and ray 2 have a λ/2 shift upon reflection. Thus, to obtain destructive interference, ray 2 needs to travel a half wavelength farther than ray 1. For rays incident perpendicularly, the path length difference is 2t.

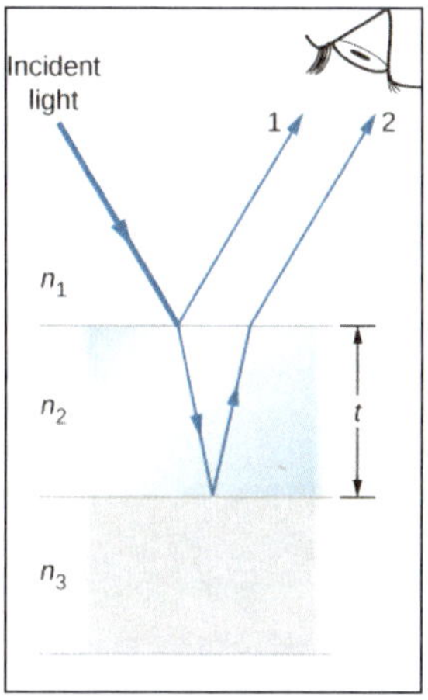

To obtain destructive interference here,

$$2t = \frac{\lambda_{n2}}{2}$$

Where λ_{n2} is the wavelength in the film and is given by $\lambda_{n2} = \lambda/n_2$. Thus,

$$2t = \frac{\lambda / n_2}{2}.$$

Solving for t and entering known values yields

$$t = \frac{\lambda / n_2}{4} = \frac{(500\ nm)/1.38}{4} = 99.6\ nm.$$

Significance

Films such as the one in this example are most effective in producing destructive interference when the thinnest layer is used, since light over a broader range of incident angles is reduced in intensity. These films are called nonreflective coatings; this is only an approximately correct description, though, since other wavelengths are only partially cancelled. Nonreflective coatings are also used in car windows and sunglasses.

Combining Path Length Difference with Phase Change

Thin-film interference is most constructive or most destructive when the path length difference for the two rays is an integral or half-integral wavelength. That is, for rays incident perpendicularly,

$$2t = \lambda_n, 2\lambda_n, 3\lambda_n, ... or \; 2t = \lambda_n / 2, 3\lambda_n / 2, 5\lambda_n / 2, ...$$

To know whether interference is constructive or destructive, you must also determine if there is a phase change upon reflection. Thin-film interference thus depends on film thickness, the wavelength of light, and the refractive indices. For white light incident on a film that varies in thickness, you can observe rainbow colors of constructive interference for various wavelengths as the thickness varies.

Example: Soap Bubbles

- What are the three smallest thicknesses of a soap bubble that produce constructive interference for red light with a wavelength of 650 nm? The index of refraction of soap is taken to be the same as that of water.
- What three smallest thicknesses give destructive interference?

Strategy

Use Figure given below to visualize the bubble, which acts as a thin film between two layers of air. Thus $n_1 = n_3 = 1.00$ for air, and $n_2 = 1.333$ for soap (equivalent to water). There is a $\lambda/2$ shift for ray 1 reflected from the top surface of the bubble and no shift for ray 2 reflected from the bottom surface. To get constructive interference, then, the path length difference (2t) must be a half-integral multiple of the wavelength—the first three being $\lambda_n/2$, $3\lambda_n/2$, and $5\lambda_n/2$. To get destructive interference, the path length difference must be an integral multiple of the wavelength—the first three being 0, λ_n, and $2\lambda_n$.

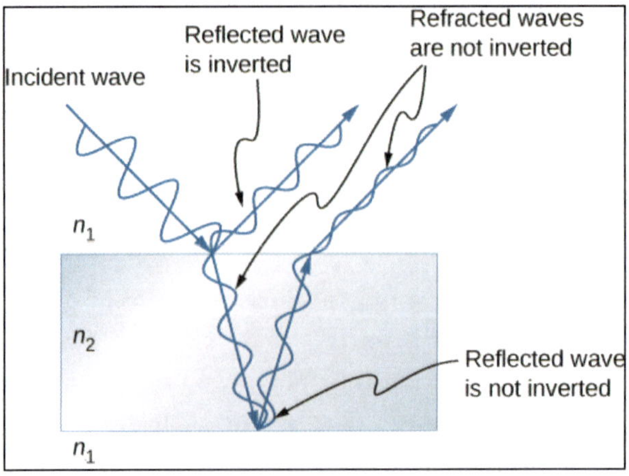

a. Constructive interference occurs here when,

$$2t_c = \frac{\lambda_n}{2}, \frac{3\lambda_n}{2}, \frac{5\lambda_n}{2}, \ldots$$

Thus, the smallest constructive thickness t_c is,

$$t_c = \frac{\lambda_n}{4} = \frac{\lambda / n}{4} = \frac{(650\ nm)/1.333}{4} = 122\ nm.$$

The next thickness that gives constructive interference is $t_c^{'} = 3\lambda_n / 4$, so that,

$$t_c^{'} = 366\ nm.$$

Finally, the third thickness producing constructive interference is $t_c^{'} = 5\lambda_n / 4$, so that,

$$t_c^{'} = 610\, nm.$$

b. For destructive interference, the path length difference here is an integral multiple of the wavelength. The first occurs for zero thickness, since there is a phase change at the top surface, that is,

$$t_d = 0,$$

the very thin (or negligibly thin). The first non-zero thickness producing destructive interference is,

$$2t_d^{'} = \lambda_n.$$

Substituting known values gives,

$$t_d^{'} = \frac{\lambda}{2} = \frac{\lambda / n}{2} = \frac{(650\, nm)/1.333}{2} = 244\ nm.$$

Finally, the third destructive thickness is $2t"_d = 2\lambda_n$, so that,

$$t"_d = \lambda_n = \frac{\lambda}{n} = \frac{650\ nm}{1.333} = 488\ nm.$$

Significance

If the bubble were illuminated with pure red light, we would see bright and dark bands at very uniform increases in thickness. First would be a dark band at 0 thickness, then bright at 122 nm thickness, then dark at 244 nm, bright at 366 nm, dark at 488 nm, and bright at 610 nm. If the bubble varied smoothly in thickness, like a smooth wedge, then the bands would be evenly spaced.

Another example of thin-film interference can be seen when microscope slides are separated. The slides are very flat, so that the wedge of air between them increases in thickness very uniformly. A phase change occurs at the second surface but not the first, so a dark band forms where the slides touch. The rainbow colors of constructive interference repeat, going from violet to red again and again as the distance between the slides increases. As the layer of air increases, the bands become more difficult to see, because slight changes in incident angle have greater effects on path length differences. If monochromatic light instead of white light is used, then bright and dark bands are obtained rather than repeating rainbow colors.

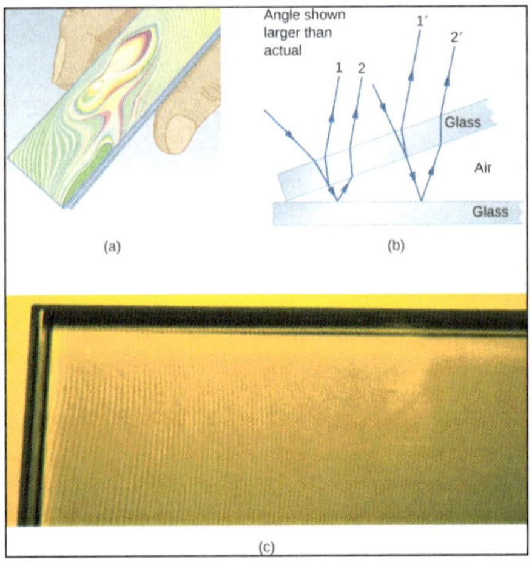

(a) The rainbow-color bands are produced by thin-film interference in the air between the two glass slides. (b) Schematic of the paths taken by rays in the wedge of air between the slides. (c) If the air wedge is illuminated with monochromatic light, bright and dark bands are obtained rather than repeating rainbow colors.

An important application of thin-film interference is found in the manufacturing of optical instruments. A lens or mirror can be compared with a master as it is being ground, allowing it to be shaped to an accuracy of less than a wavelength over its entire surface. Figure illustrates the phenomenon called Newton's rings, which occurs when the plane surfaces of two lenses are placed together. (The circular bands are called Newton's rings because Isaac Newton described them and their use in detail. Newton did not discover them; Robert Hooke did, and Newton did not believe they were due to the wave character of light.) Each successive ring of a given color indicates an increase of only half a wavelength in the distance between the lens and the blank, so that great precision can be obtained. Once the lens is perfect, no rings appear.

"Newton's rings" interference fringes are produced when two plano-convex lenses are placed together with their plane surfaces in contact. The rings are created by interference between the light reflected off the two surfaces as a result of a slight gap between them, indicating that these surfaces are not precisely plane but are slightly convex.

Thin-film interference has many other applications, both in nature and in manufacturing. The wings of certain moths and butterflies have nearly iridescent colors due to thin-film interference. In addition to pigmentation, the wing's color is affected greatly by constructive interference of certain wavelengths reflected from its film-coated surface. Some car manufacturers offer special paint jobs that use thin-film interference to produce colors that change with angle. This expensive option is based on variation of thin-film path length differences with angle. Security features on credit cards, banknotes, driving licenses, and similar items prone to forgery use thin-film interference, diffraction gratings, or holograms. As early as 1998, Australia led the way with dollar bills printed on polymer with a diffraction grating security feature, making the currency difficult to forge. Other countries, such as Canada, New Zealand, and Taiwan, are using similar technologies, while US currency includes a thin-film interference effect.

Newton's Rings

Newton's rings are the circular interference pattern first discovered by Newton.

Formation of Fringes

When a plano-convex lens with large radius of curvature is placed on a plane glass plate such that its curved surface faces the glass plate, a wedge air film (of gradually increasing thickness) is formed between the lens and the glass plate. The thickness of the air film is zero at the point of contact and gradually increases away from the point of contact.

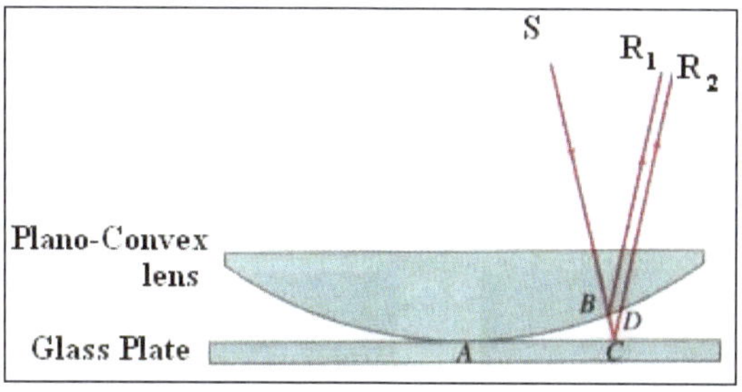

If monochromatic light is allowed to fall normally on the lens from a source 'S', then two reflected rays R_1 (reflected from upper surface of the film) and R_2 (reflected from lower surface of the air film) interfere to produce circular interference pattern. This interference pattern has concentric alternate bright and dark rings around the point of contact.

Theory of Fringes

The effective path difference between the two reflected rays R_1 and R_2 for a wedge shaped film from equation,

$$\Delta = 2\mu t\cos(r+\theta) + \frac{\lambda}{2}$$

If the light is incident normally on the lens, r = 0 and near to point of contact θ is small; therefore near point of contact, $(r+\theta) \to 0°$.

Therefore, $\Delta = 2\mu t + \frac{\lambda}{2}$.

At point of contact t = 0 therefore the effective path difference $\Delta = \lambda/2$ which is odd multiple of $\lambda/2$. Therefore the Central fringe is dark.

Condition of Maxima (Bright Fringe): The effective path difference $\Delta = \pm n\lambda$; substituting this in equation $\Delta = 2\mu t + \frac{\lambda}{2}$,

$$2\mu t + \frac{\lambda}{2} = \pm n\lambda$$

$$2\mu t = \pm\frac{(2n-1)\lambda}{2}.$$

Condition for Minima (Dark Fringe): The effective path difference $\Delta = \pm\frac{(2n+1)\lambda}{2}$; substituting this in equation $\Delta = 2\mu t + \frac{\lambda}{2}$,

$$2\mu t + \frac{\lambda}{2} = \pm\frac{(2n+1)\lambda}{2}$$

$$2\mu t = \pm n\lambda$$

From $2\mu t = \pm\frac{(2n-1)}{2}$ and $2\mu t = \pm n\lambda$ it is clear that for particular dark or bright fringe t should be constant.

Every fringe is the locus of points having equal thickness. Hence the fringes are circular in shape.

Diameter of Newton's Rings

To calculate the diameter of fringes, assume a plano-convex lens is placed on a plane glass plate as shown in figure say R be the radius of curvature of lens. In ΔO'ML

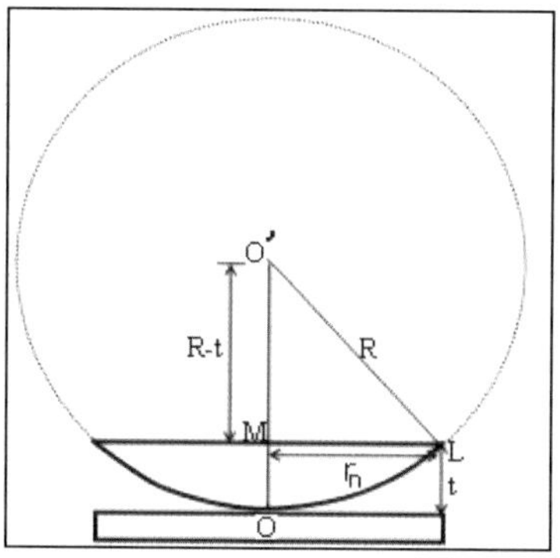

$$(OL)^2 = (O'M)^2 + (ML)^2$$
$$R^2 = (R-t)^2 + r_n^2$$
$$R^2 = R^2 + t^2 - 2Rt + r_n^2$$
$$r_n^2 = 2Rt - t^2$$

As t << R; neglecting t^2 (small value),

$$r_n^2 = 2Rt$$
$$t = \frac{r_n^2}{2R}.$$

Diameter of Dark Rings

Substitute t from equation $t = \frac{r_n^2}{2R}$ to equation $2\mu t = \pm n\lambda$,

$$2\mu \frac{r_n^2}{2R} = n\lambda$$
$$r_n^2 = n\lambda R / \mu$$

As $D_n = 2r_n$

$$D_n^2 = 4n\lambda R / \mu$$
$$D_n = \sqrt{4n\lambda R / \mu}$$

The medium enclosed between the lens and glass plate is if air therefore, $\mu = 1$. The diameter of nth order dark fringe will be,

$$D_n = \sqrt{4n\lambda R}$$
$$D_n \alpha \sqrt{n}; \quad n = 0,1,2,3,4...$$

The diameter of dark ring is proportional to square root of natural numbers.

Diameter of Bright Rings

Substitute t from equation $t = \frac{r_n^2}{2R}$ to equation $2\mu t = \pm n\lambda$

$$2\mu.\frac{r_n^2}{2R} = \frac{(2n-1)\lambda}{2}$$

$$r_n^2 = \frac{(2n-1)\lambda R}{2\mu}$$

As $D_n = 2r_n$

$$D_n^2 = 2(2n-1)\lambda R / \mu$$

$$D_n = \sqrt{2(2n-1)\lambda R / \mu}$$

The medium enclosed between the lens and glass plate is if air therefore, = 1. The diameter of nth order bright fringe will be,

$$D_n = \sqrt{2(2n-1)\lambda R}$$

$$D_n \alpha \sqrt{(2n-1)}; n = 0,1,2,3,4....$$

The diameter of bright ring is proportional to square root of odd natural numbers

Spacing between Fringes

The Newton's rings are not equally spaced because the diameter of ring does not increase in the same proportion as the order of ring and rings get closer and closer as 'n' increases.

For example the diameter of dark ring is given by $D_n = \sqrt{4n\lambda R}$

Where n = 0, 1, 2, 3, 4....

$$D_5 - D_4 = \left(\sqrt{5} - \sqrt{4}\right)\sqrt{\lambda R} = 0.236\sqrt{\lambda R}$$

$$D_{25} - D_{24} = \left(\sqrt{25} - \sqrt{24}\right)\sqrt{\lambda R} = 0.101\sqrt{\lambda R}$$

$$D_{15} - D_{14} = \left(\sqrt{15} - \sqrt{14}\right)\sqrt{\lambda R} = 0.131\sqrt{\lambda R}$$

From above, we conclude that the fringe width reduces with increase in 'n'.

Newton's Ring with White Light

If the monochromatic source is replaced by the white light few coloured rings are seen around dark centre later illumination is seen in the field of view.

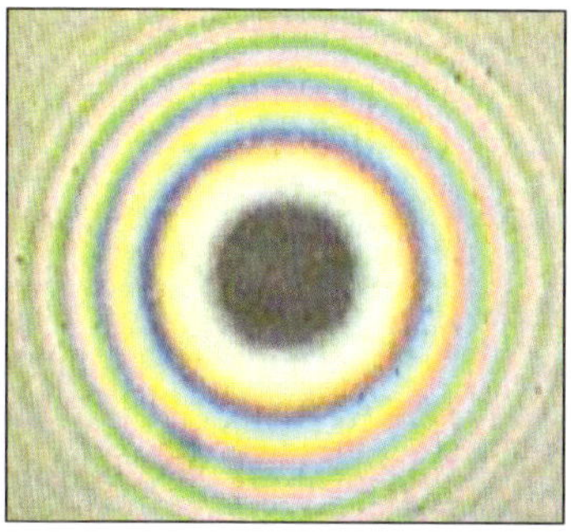

The Michelson Interferometer

The Michelson interferometer is a precision instrument that produces interference fringes by splitting a light beam into two parts and then recombining them after they have traveled different optical paths. Figure depicts the interferometer and the path of a light beam from a single point on the extended source S, which is a ground-glass plate that diffuses the light from a monochromatic lamp of wavelength λ_0. The beam strikes the half-silvered mirror M, where half of it is reflected to the side and half passes through the mirror. The reflected light travels to the movable plane mirror M_1, where it is reflected back through M to the observer. The transmitted half of the original beam is reflected back by the stationary mirror M_2 and then toward the observer by M.

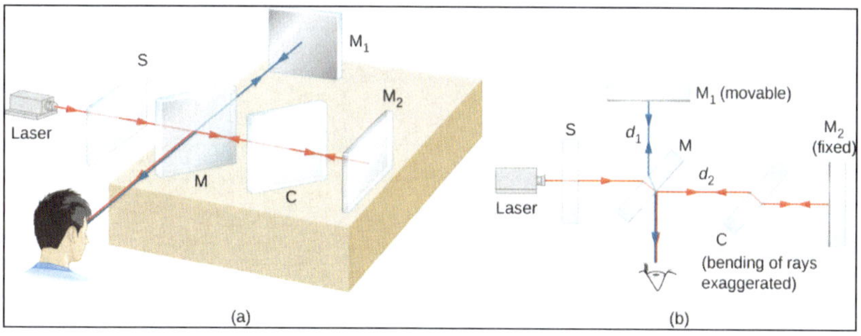

(a) The Michelson interferometer. The extended light source is a ground-glass plate that diffuses the light from a laser. (b) A planar view of the interferometer.

Because both beams originate from the same point on the source, they are coherent and therefore interfere. Notice from the figure that one beam passes through M three times and the other only once. To ensure that both beams traverse the same thickness of glass, a compensator plate C of transparent glass is placed in the arm containing M_2. This plate is a duplicate of M (without the silvering) and is usually cut from the same piece of glass used to produce M. With the compensator in place, any phase difference between the two beams is due solely to the difference in the distances they travel.

The path difference of the two beams when they recombine is $2d_1 - 2d_2$, where d_1 is the distance between M and M_1, and d_2 is the distance between M and M_2. Suppose this path difference is an integer number of wavelengths $m\lambda_0$. Then, constructive interference occurs and a bright image of the point on the source is seen at the observer. Now the light from any other point on the source whose two beams have this same path difference also undergoes constructive interference and produces a bright image. The collection of these point images is a bright fringe corresponding to a path difference of $m\lambda_0$. When M_1 is moved a distance $\Delta d = \lambda_0/2$, this path difference changes by λ_0, and each fringe moves to the position previously occupied by an adjacent fringe. Consequently, by counting the number of fringes m passing a given point as M_1 is moved, an observer can measure minute displacements that are accurate to a fraction of a wavelength, as shown by the relation,

$$\Delta d = m\frac{\lambda_0}{2}.$$

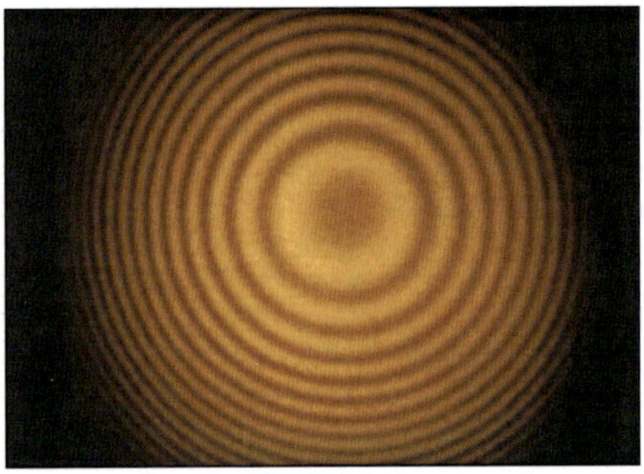

Fringes produced with a Michelson interferometer.

Precise Distance Measurements by Michelson Interferometer

A red laser light of wavelength 630 nm is used in a Michelson interferometer. While keeping the mirror M_1 fixed, mirror M_2 is moved. The fringes are found to move past a fixed cross-hair in the viewer. Find the distance the mirror M_2 is moved for a single fringe to move past the reference line.

Strategy

We use the result of the Michelson interferometer interference condition to find the distance moved, Δd.

For a 630-nm red laser light, and for each fringe crossing ($m = 1$), the distance travelled by M_2 if you keep M_1 fixed is,

$$\Delta d = m\frac{\lambda_0}{2} = 1\times\frac{630\ nm}{2} = 315\ nm = 0.315\mu m.$$

Significance

An important application of this measurement is the definition of the standard meter. The length of the standard meter was once defined as the mirror displacement in a Michelson interferometer corresponding to 1,650,763.73 wavelengths of the particular fringe of krypton-86 in a gas discharge tube.

Measuring the Refractive Index of a Gas

In one arm of a Michelson interferometer, a glass chamber is placed with attachments for evacuating the inside and putting gases in it. The space inside the container is 2 cm wide. Initially, the container is empty. As gas is slowly let into the chamber, you observe that dark fringes move past a reference line in the field of observation. By the time the chamber is filled to the desired pressure, you have counted 122 fringes move past the reference line. The wavelength of the light used is 632.8 nm. What is the refractive index of this gas?

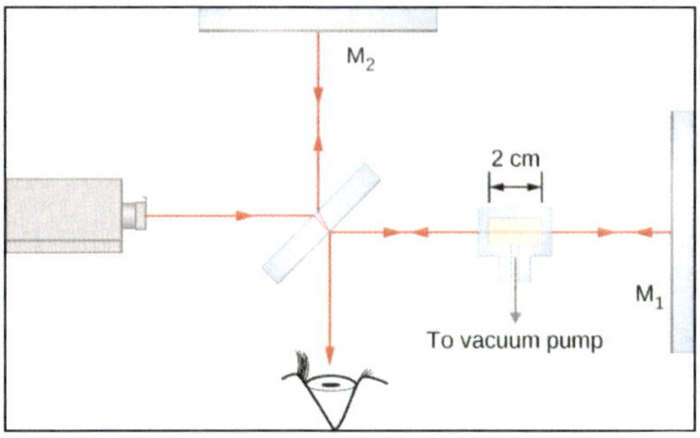

Strategy

The m = 122 fringes observed compose the difference between the number of wavelengths that fit within the empty chamber (vacuum) and the number of wavelengths that fit within the same chamber when it is gas-filled. The wavelength in the filled chamber is shorter by a factor of n, the index of refraction.

The ray travels a distance t = 2 cm to the right through the glass chamber and another distance t to the left upon reflection. The total travel is L = 2t. When empty, the number of wavelengths that fit in this chamber is,

$$N_0 = \frac{L}{\lambda_0} = \frac{2t}{\lambda_0}$$

Where λ_0 = 632.8 nm is the wavelength in vacuum of the light used. In any other medium, the wavelength is $\lambda = \lambda_0/n$ and the number of wavelengths that fit in the gas-filled chamber is,

$$N = \frac{L}{\lambda} = \frac{2t}{\lambda_0 / n}.$$

The number of fringes observed in the transition is,

$$m = N - N$$

$$= \frac{2t}{\lambda_0} - \frac{2t}{\lambda_0}$$

$$= —(\quad -1).$$

Solving for (n – 1) gives,

$$n - 1 = m\left(\frac{\lambda_0}{2t}\right) = 122\left(\frac{632.8 \times 10^{-9}m}{2\left(2 \times 10^{-2}m\right)}\right) = 0.0019$$

and n = 1.0019.

Significance

The indices of refraction for gases are so close to that of vacuum, that we normally consider them equal to 1. The difference between 1 and 1.0019 is so small that measuring it requires a correspondingly sensitive technique such as interferometry. We cannot, for example, hope to measure this value using techniques based simply on Snell's law.

Applications

The key applications of a Michelson interferometer are as follows:

- In the Michelson-Morley experiment, which led to the development of the special theory of relativity. This was a failed attempt to demonstrate a hypothetical "aether wind" as a theory of light.
- The detection of gravitational waves – LIGO is a massive interferometer with two lasers positioned thousands of kilometres apart.
- In astronomical interferometry.
- In optical coherence tomography.
- In fibre optics.
- In analysing the upper atmosphere, by revealing temperatures and winds, and by measuring the Doppler widths and shifts in the spectra of airglow and aurora.
- As a component in the helioseismic and magnetic imager to study solar variability, and to illustrate the Sun's interior, along with the many aspects of magnetic activity.
- As a tunable narrow band filter.
- As the core of Fourier transform spectroscopy.

Multiple-beam Interferometry

The technique of multiple-beam interferometry is based upon situating two surfaces of high reflectivity in close proximity and using a lens to converge beams which have undergone multiple-reflection between the surfaces.

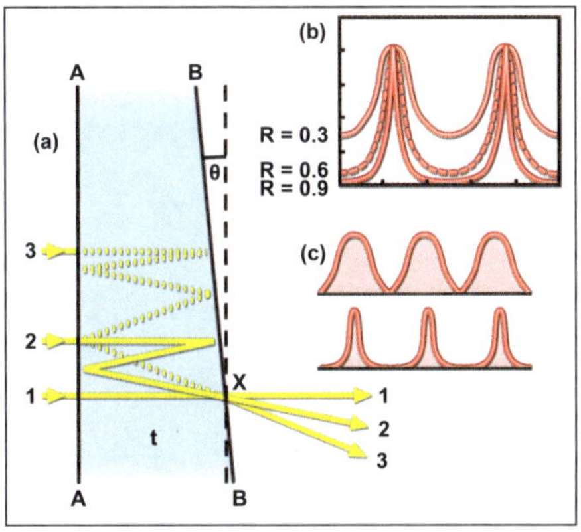

Multiple-beam Reflections and Interference.

An arrangement in which the two opposed surfaces are parallel is utilized in the Fabry-Perot interferometer. If the two planes are not parallel, then interference fringes appear localized in the wedge space. The situation is essentially the same as that encountered in the wedge space in two-beam interferometry.

Characteristics of Multiple-beam Interference Patterns

In multiple-beam interferometry, the breadth of the interference fringes becomes extremely narrow. That is, the contour lines on the "map" become narrow and precision of measurement of surface topography improves accordingly. When the optimal conditions are satisfied, the breadth of the fringes in multiple-beam interferometry is on the order of 1/50 of the corresponding breadth in the two-beam method, and hence the precision is improved by a factor of approximately 50. Because, the limit of measurement of level differences utilizing that technique is about 25 nanometers, the limit of measurement in multiple-beam interferometry is therefore on the order of 0.5 nanometer (5 Angstroms).

As shown in figure, when the incident light enters the wedge space, multiple-reflection between the opposing surfaces occurs. The beams arriving at point x in the figure include the non-reflected beam (1), the twice-reflected beam (2), the fourfold-reflected beam (3), etc. At each reflection, the intensity drops in accordance with the reflectivity of the surface, and therefore the quantity of light of multiple-beams finally collected by the lens is governed by the reflectivity. The greater the number of multiple reflections of the beams contributing to the interference, the finer and sharper the fringes. Figure illustrates the relation between the breadth of the interference fringes and the reflectivity for the case in which the reflectivities of the specimen surface and the reference plate (reference mirror) are identical. It can be seen from the figure that the reflectivity determines the breadth of the interference fringes.

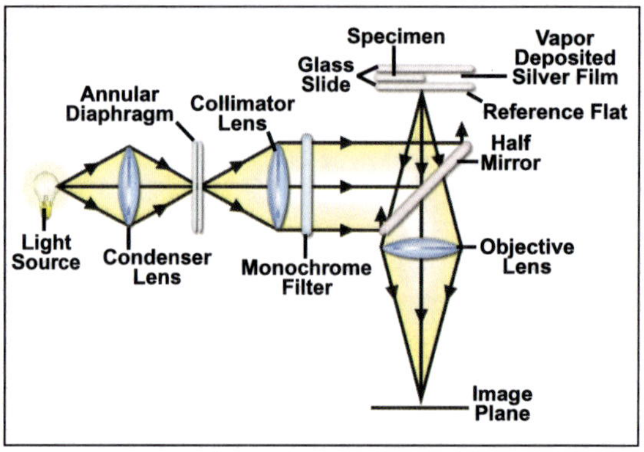

Low Magnification Objective System.

A comparison between the intensity distributions of two-beam interference fringes and multiple-beam interference fringes corresponding to 90 percent reflectivity is shown in figure. The two-beam interference fringes are of the form (cosθ)E2, and the breadth of the ridges and valleys is nearly the same. On the other hand, as shown in the lower curve of figure, the multiple-beam interference fringes display a sharply peaked intensity distribution. As in the case of two-beam interferometry, successive interference fringes appear localized at every level change of half a wavelength. However, because the fringes are narrow, information concerning the regions between the fringes is likely to be unavailable.

Conditions for High Precision Measurement

Proceeding from the Airy formula for multiple reflection between parallel plane surfaces, Samuel Tolansky performed a detailed analysis of the intensity distribution of the multiple-interference fringes appearing in a wedge space, and from the results of this study deduced the conditions for obtaining the highest measurement precision, as summarized by the following five points:

- Coating the surface of the reference plate with a film of high reflectivity and low absorption.
- Coating the specimen with a uniform film of high reflectivity, faithfully conforming with the original topography of the specimen.
- Making the distance between the two surfaces (t in figure) as small as possible, at most 10 millimeters, and preferably of the order of the wavelength of the light.
- Collimating the incident light beam to a parallelism within 3 degrees.
- Making the incident light as closely perpendicular to the reference plate as possible.

The last two conditions are comparatively easy to achieve provided that the previous one is satisfied. In summary, the surfaces of the reference plate and the specimen should be brought into close proximity, reducing as far as possible the distance t in figure, and the surface of the reference mirror should be coated by silver vapor deposition or with a multi-layer film of low absorption.

Optical Systems for Multiple-beam Interferometry

The basic function of optical systems for multiple-beam interferometry is to allow observation of the interference pattern appearing in the wedge space formed by two surfaces. Observation of the fringes produced by reflection can be performed with a microscope in two possible ways. Figure illustrates an optical system appropriate for the measurement of film thickness, etc., utilizing a low power objective lens. Because the working distance of the low power objective is relatively long, a half mirror can be inserted between the objective and the specimen. Additionally, the low power objective possesses the advantage of a large depth of field, which permits observation over a wide field of view.

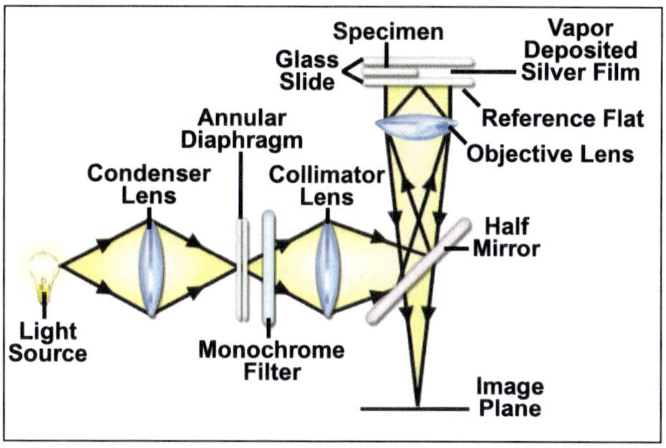

High Magnification Objective System.

Although the low power lens has certain benefits, a high magnification objective lens is necessary for measurement of the minute topography and relief features of the specimen. With such an objective, because the working distance is reduced to less than 1 millimeter, an optical system of the type depicted in figure must be utilized. This is the optical system of an ordinary reflecting microscope. It is ideal to converge the light at the rear focal point of the objective, so that the light beam is directed perpendicularly to the specimen surface. Commercially available objective lenses are tolerably suitable for this purpose.

Low-pressure mercury lamps of the type used for spectroscopy were formerly employed as light sources in such interferometric systems. Using this type of lamp, sharp spectral lines can be obtained, and additionally, the spectral arrangement of colors can be used to identify the sequential order of the interference fringes. The disadvantage of these lamps is low illumination intensity.

Recently, however, bright halogen lamps combined with interference filters have come into general use for interferometry. With achromatic objectives, green (546-nanometer) monochromatic light is the least expensive type of illumination, and provides excellent results.

Specimen Preparation

Adequate preparation of the specimen is essential in order to obtain the highest precision in multiple-beam interferometry. Because level differences of atomic dimensions are to be measured, contaminants must be removed from the surface. Therefore, the surface should be thoroughly cleaned, except for cases where the surface is mechanically weak or the specimen is chemically unstable. The same requirements apply to cleaning of the reference flat. Commercially marketed multiple-beam interferometric sets provide a selection of multilayer optical flat with various reflectivities matched to the reflectivity of specimens (cleaning of these plates may erode the coating and therefore requires great care). Use of flats matched to the specimen reflectivity permits the formation of multiple-beam interference fringes, but, as illustrated in figure, the sharpness of the fringes depends upon the reflectivities of the specimen and the reference plate. Consequently, imparting a high reflectivity to both surfaces is the most effective means of ensuring the formation of a distinct pattern.

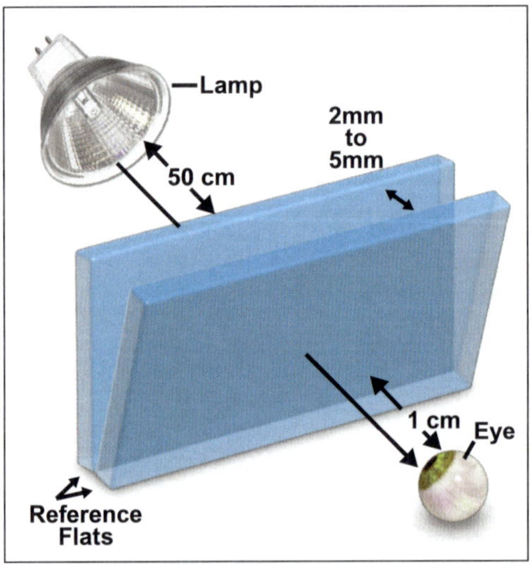

Reference Flat Evaluation.

Especially small specimens should be cleaned after first being mounted on a glass slide with Canada balsam or another mounting medium. Chemically and mechanically robust materials such as glass and quartz should be cleaned by the following procedure:

- Apply a small quantity of a suitable surfactant (such as a detergent of the type used for kitchen utensils) to a wad of absorbent cotton, and remove the dirt from the specimen by vigorous scrubbing. Ultrasonic cleaning is also suitable for this purpose. However, grease is not easily removed without scrubbing.

- Wet a wad of absorbent cotton with hydrogen peroxide and clean in the same manner as described in the step above.
- Clean the specimen in the same manner with distilled water.
- Wipe the surface thoroughly with dry absorbent cotton until the fog disappears instantly after breathing onto the surface.

Specimens with low reflectivity should be coated by vacuum deposition with a material such as aluminum, silver, or gold. Silver is particularly easy to apply, and is of high reflectivity.

A vacuum-deposited silver film having a thickness of 50-100 nanometers is appropriate. Coating should be performed by rapid vapor deposition so as to be completed in 20 to 30 seconds. In order to accomplish this, a sufficient quantity of silver is placed in a vapor deposition boat, and a shutter is interposed between the specimen and the boat. After the silver is white hot, the shutter is opened and then closed, providing a simple means of controlling the thickness of the coating by varying the time that the shutter is open. In this setup, the distance between the specimen and the boat should be at least 20-30 centimeters.

If the thickness of the deposited silver film is roughly 50-100 nanometers, then the film will appear bluish violet when viewed against a bright light source. An excessively thick film will have a surface that behaves like a mirror, and will not transmit light.

The quality of the silver film on the reference plate can be evaluated as follows. If the silvered surfaces of two such plates are opposed and subtend a wedge as illustrated in figure, and a bright light source is viewed through the wedge, then a sequence of similar images of the light source will appear. The number of these images should be counted; the presence of 25 or more indicates a satisfactory vapor-deposited film, having a reflectance of at least 90 percent. As a rough criterion, the more slowly the color of the images shifts from blue to red, the better the quality of the film.

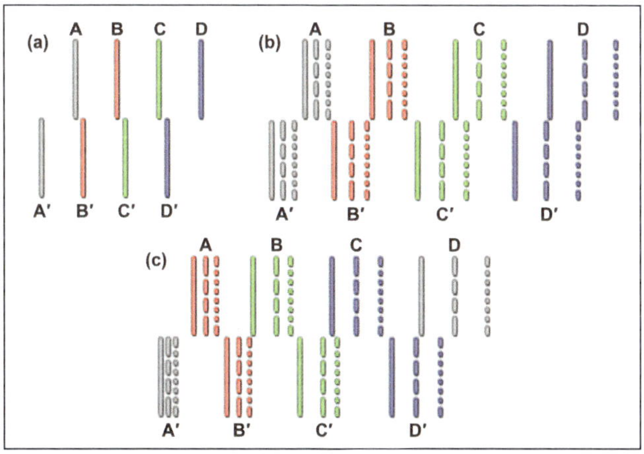

Recognition of Interference Fringe Order.

Practical Applications of Multiple-beam Interferometry

When surface topography is measured by multiple-beam interferometry, the specimen and the reference plate make contact at some point. Hence, this method is not appropriate for specimens that are sensitive to contamination or for very soft specimens. Moreover, specimens which for some reason cannot be coated with silver are not suitable.

For sharpness of interference fringes, the closest possible proximity of the specimen and the reference plate is essential. In order to realize this, the specimen and reference flat are clamped in a jig having three screws that must be adjusted while observing the interference fringes through the microscope. As in the case of two-beam interferometry, the pattern must be adjusted so that the fringes appear perpendicular to the step to be measured, and the dispersion of the fringes is also adjusted. Measurement can be performed if at least three fringes appear in the view field.

The continuity of the interference fringes can be ascertained from the color arrangement of the fringes formed under illumination by white light, as in two-beam interferometry. Examples are presented in figure that illustrate the reasoning involved. Suppose that the interference pattern is observed using a 546-nanometer filter with a low-pressure mercury lamp of the type employed for spectroscopy, and that a shift of the fringes appears as shown in figure. Obviously, the upper and lower rows of fringes have been mutually displaced by a step on the surface, but the original manner of continuity of the fringes is not evident. If the monochromatic filter is removed, interference fringes with yellow and orange spectral fringes appear, and the proper connection of the fringes can be ascertained by comparing the arrangement of the colored fringes. In figure, B-B' represents a sequence of interference fringes of the same order, while in figure, the fact that B-C' represents interference fringes of the same order is equally obvious by inspection of the arrangement of the colored fringes.

However, in the case of multiple-beam interferometry, the spacing of the interference fringes is not necessarily equidistant, and therefore height is calculated by the following method. In figure, for example, the height, h, of a step can be determined by the following formula:

$$h = 2BB'/(A'B' + B'C') \times \lambda/2 = BB'/(A'B' + B'C') \times \lambda.$$

Likewise, for the example illustrated in figure, the appropriate formula is:

$$h = 2BC'/(B'C' + B'C') \times \lambda/2 = BC'/(B'C' + B'C') \times \lambda.$$

A multiple-beam interferometric photomicrograph of the pits, termed trigons (triangular depressions), in a natural diamond is presented in figure. The larger depressions are 10-20 nanometers in depth, while the depths of the smaller ones are on the order of 2-4 nanometers.

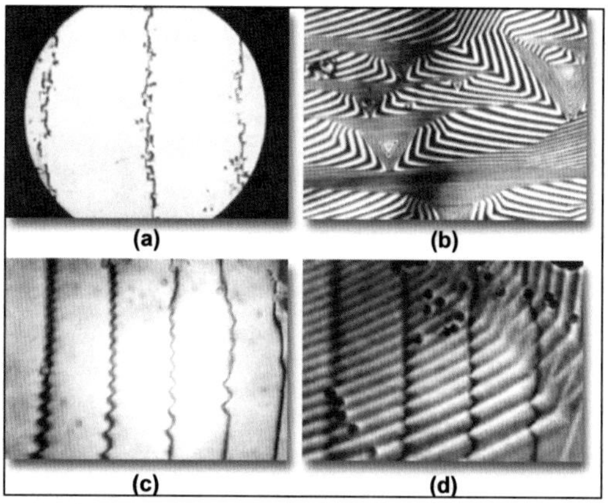

Multiple-Beam Interference Measurements.

Figure illustrates a multiple-beam interference pattern, formed by trigons, utilizing transmitted light, which can be observed in this case since the diamond itself transmits light. Transmitted multiple-beam interference displays bright fringes. This contrasts with patterns formed by opaque specimens, which show dark fringes. Light transmitted through specimens which possess double refractive indices forms exceedingly complicated interference patterns and therefore such specimens are not suitable for measurement by means of transmitted light. In such cases, reliable measurements of topographic irregularities can still be obtained by interference of reflected light.

High dispersion interference permits the utilization of intensity variations due to extremely minute differences in height. This is illustrated by the example of the trigons of a natural diamond shown in figure. The surface shown in this micrograph is nearly the same as that shown in figure, but in that instance, the surface of the specimen and the reference plate subtend a wedge, so that a small number of interference fringes are formed. However, in the formation of the pattern of figure, the two surfaces are positioned almost in parallel, thereby spreading a single fringe over a large area and facilitating detailed observation. The contrast displayed by minute level changes can be enhanced to an even greater extent than is possible when utilizing two-beam interferometry. As shown in figure, an elevation difference dt gives rise to an intensity difference dI. The gradient of the intensity distribution of a multiple-beam interference fringe is far more abrupt than that of a two-beam interference fringe, and the contrast due to minute level variations displays correspondingly greater sensitivity.

A hexagonal ferrite crystal (magnetoplumbite) is presented in figure. Here, figure is a portion of the multiple-beam interference pattern photographed at high magnification. Figure is a phase contrast photomicrograph of the same pattern portion, superimposed upon an interference pattern. In this manner, level differences can be displayed quantitatively on a single photograph. Moreover, one can compensate for the loss of information in the intermediate regions between fringes that results from excessive spacing between the fringes in the multiple-beam interference pattern.

Fabry-perot Interferometer

The Fabry-Perot interferometer is an optical instrument which uses multiple-beam interference.

Interference at a Plane-parallel Plate

First we calculate the transmitted and reflected intensity when shining a plane wave of monochromatic light on a plane-parallel plate.

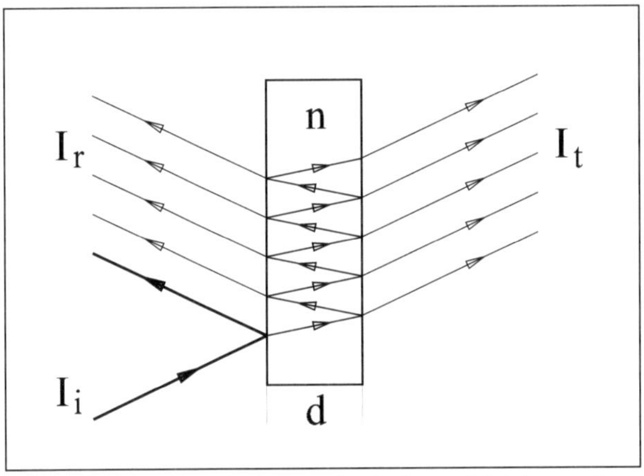

Interference at a plane-parallel plate of thickness d.

The situation is depicted in figure, light of intensity I_i, wavelength λ and wavevector $k = 2\pi / \lambda$ is incident on a plate of thickness d and refractive index n. The two surfaces affected may both be characterized by a reflectance R and a transmittance T, defined for the light intensities. The absorption of the plate material should be negligible, thus R + T = 1 (energy conservation).

To calculate the transmitted and reflected intensities I_t and I_r, we have to sum up the amplitudes of the respective waves, taking into account the phase shifts imposed by the plate. Reflectance and transmittance for the amplitudes at the plate interfaces, r, r', t, and t', respectively, are defined according to figure.

Relations between the reflectance and transmittance values for amplitudes can be derived using time reversal invariance. If losses due to absorption can be neglected, the propagation of waves must be reversible. Figure shows the 'normal' and the reversed beam propagation through a surface and denotes the corresponding amplitudes.

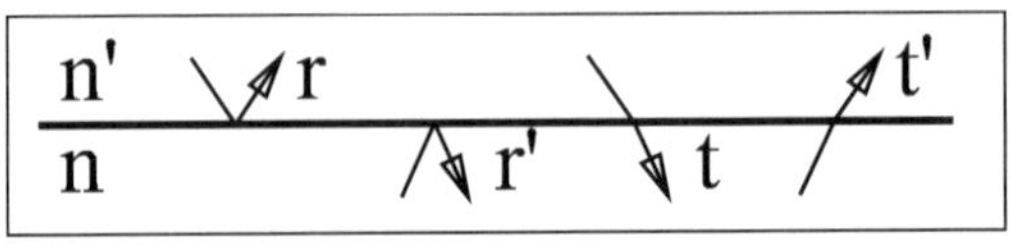

Definition of reflectance and transmittance at an interface between two nonabsorbing materials.

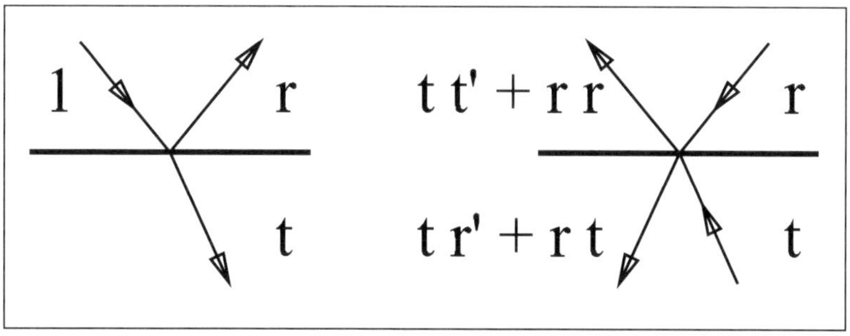

'Normal' (left) and time reversed (right) reflection and transmission at an interface.

A comparison yields,

$$tt' + rr = 1$$
$$tr' + rt = 0$$

Therefore,

$$r' = -r, \quad r'^2 = r^2 = R, \quad tt' = T = 1 - R.$$

Figure, to yield a descriptive visualization, is sketched with oblique incidence of the light – the derivation of the transfer function, yet, will be done using normal incidence.

The reflected amplitude is calculated by summing all reflected waves. Different waves m differ in phase due to the optical path difference 2mdn and in amplitude due to reflection and transmission by $tt' \cdot r'^{2m-1}$. The first reflected wave has to be considered with special care, here the reflectance for the 'out-in' direct $-r-$ has to be used. Putting all together, we get for the reflected amplitude.

$$A_r = A_i r = A_i \exp(jk2dn)\, tt' r' \sum_{m=0}^{\infty} r'^{2m} \exp(jk2mdn)$$

$$= A_i r - A_i \exp(j\delta)\, Tr \sum_{m=0}^{\infty} \left(R \exp(j\delta)\right)^m, \quad \delta = 2kdn$$

$$= A_i r - A_i \exp(j\delta)\, T_r \frac{1}{1 - R\exp(j\delta)}$$

$$= A_i r \frac{1 - R\exp(j\delta) - T\exp(j\delta)}{1 - R\exp(j\delta)} = A_i r \frac{1 - \exp(j\delta)}{1 - \mathrm{R}\exp(j\delta)}$$

And for the intensity,

$$I_r = A_r \cdot A_r^*$$

Which – after some conversions – yields,

$$\frac{I_r}{I_i}=\frac{2R(1-\cos\delta)}{1-2R\cos\delta+R^2}=\frac{4R\sin^2\frac{\delta}{2}}{(1-R)^2+4R\sin^2\frac{\delta}{2}}.$$

For the transmitted amplitude the expression gets a little bit simpler,

$$\begin{aligned}A_t&=A_i tt'\sum_{m=0}^{\infty}r'^{2m}\exp(jk2mdn)\\&=A_i T\sum_{m=0}^{\infty}R^m\exp(j\delta)\\&=A_i\frac{T}{1+\mathrm{R}\exp(j\delta)}\end{aligned}$$

And,

$$\frac{I_t}{I_i}=\frac{T^2}{1-2R\ \cos\delta+R^2}=\frac{T^2}{(1-R)^2+4R\sin^2\frac{\delta}{2}}.$$

Equation $\frac{I_t}{I_i}=\frac{T^2}{1-2R\ \cos\delta+R^2}=\frac{T^2}{(1-R)^2+4R\sin^2\frac{\delta}{2}}.$ of course also may be obtained directly from Equation $\frac{I_r}{I_i}=\frac{2R(1-\cos\delta)}{1-2R\cos\delta+R^2}=\frac{4R\sin^2\frac{\delta}{2}}{(1-R)^2+4R\sin^2\frac{\delta}{2}}.$ $=I_r+I_t=I_i=$ I_i (energy conservation law). The expressions for the reflected and transmitted intensities Eqs. $\frac{I_r}{I_i}=\frac{2R(1-\cos\delta)}{1-2R\cos\delta+R^2}=\frac{4R\sin^2\frac{\delta}{2}}{(1-R)^2+4R\sin^2\frac{\delta}{2}}.$ and $\frac{I_t}{I_i}=\frac{T^2}{1-2R\ \cos\delta+R^2}=\frac{T^2}{(1-R)^2+4R\sin^2\frac{\delta}{2}}.$ are known as Airy's formula.

One can define a quality factor for the reflect,

$$Q_R=\frac{4R}{(1-R)^2}$$

which simplifies the expressions Eqs. $\frac{I_r}{I_i}=\frac{2R(1-\cos\delta)}{1-2R\cos\delta+R^2}=\frac{4R\sin^2\frac{\delta}{2}}{(1-R)^2+4R\sin^2\frac{\delta}{2}}$ and

$\frac{I_t}{I_i}=\frac{T^2}{1-2R\ \cos\delta+R^2}=\frac{T^2}{(1-R)^2+4R\sin^2\frac{\delta}{2}}$ further,

$$\frac{I_r}{I_i}=\frac{Q_R\sin^2\frac{\delta}{2}}{1+Q_R\sin^2\frac{\delta}{2}}$$

$$\frac{I_t}{I_i}=\frac{1}{1+Q_R\sin^2\frac{\delta}{2}}$$

This quality factor Q_R sometimes is denoted as finesse coefficient.

It should be noted that at the transmission peaks all of the incident intensity is transmitted, nothing reflected, even when using highly reflecting surfaces.

Figure shows a plot of the transmittance function equation $\frac{I_t}{I_i}=\frac{1}{1+Q_R\sin^2\frac{\delta}{2}}$. for different values of Q_R. Instead of δ, the corresponding interference $N=\delta/2\pi$ is noted.

Two properties are characteristic for each of the transmission functions, the width of the maxima and the suppression of light throughput between the maxima.

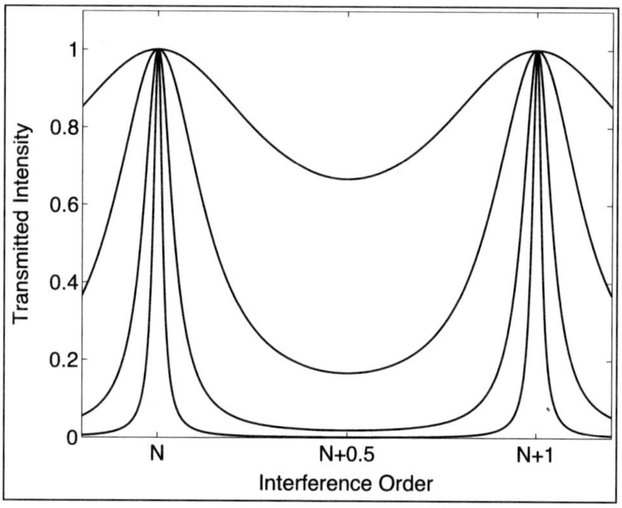

The transfer function I_t/I_i of a plane-parallel plate (Fabry-Perot plate) for different reflectivities (quality factors 0.5, 5, 50, and 500, respectively, corresponding to reflectivities R of 0.1, 0.42, 0.75, 0.91).

Applications

The transfer function shows high transmission when certain conditions for the wavelength of the light and the plate geometry are fulfilled, otherwise the light is more or less blocked. These conditions can be adjusted by changing the wavelength and/or the geometry. Thus Fabry-Perot plates can be used to measure or to control light wavelengths or to measure geometric properties. The geometric conditions are defined by several properties including thickness, refractive index, and beam direction. As a Fabry-Perot arrangement is very sensitive with regard to the geometry, small deviations in these properties can be measured. The same sensitivity facilitates various optical applications. A Fabry-Perot plate, e. g., can be used in a laser resonator to select single longitudinal modes, the adjustment is made by tilting it with respect to the beam direction.

Very important is the application for high-resolution optical spectroscopy. To use a Fabry-Perot interferometer there, one of the geometric properties has to be adjustable and to be related in a suitable way to wavelength variations. With plates, this can be achieved when using divergent light, thus different beam directions. More customary are air-spaced interferometers, where spacing or air-pressure are used for tuning.

Fringes of Equal Inclination

Light source for a Fabry-Perot plate either an extended light source in the focal plane of a collimating lens or a divergent point source produces concentric circular interference fringes due to the inclination dependent phase shifts (fringes of equal inclination).

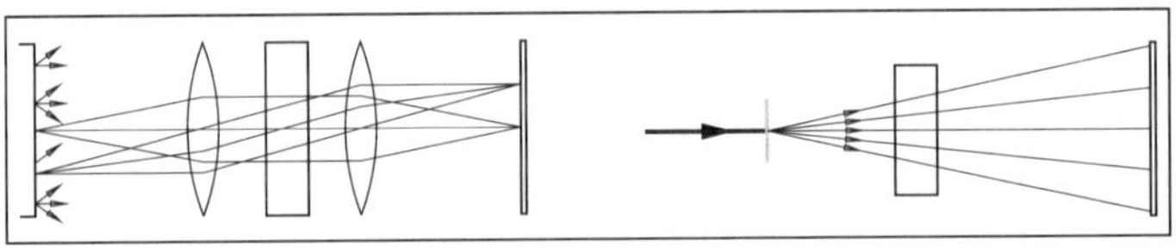

Experimental arrangements for fringes of equal inclination. Left: extended source in the focal plane of the collimating lens, right: divergent point source.

Typical fringe patterns for monochromatic light sources (calculated for plates of different reflectivity) are shown in figure.

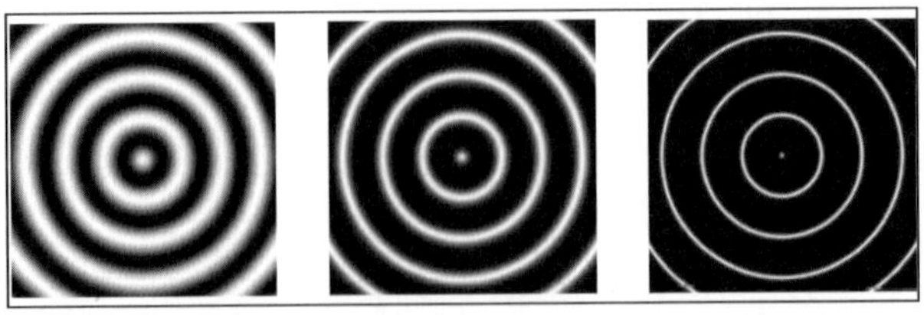

Fringes of equal inclination for Fabry-Perot plates of different reflectivity. From left to right: finesse coefficients = 1, 10, 100.

Arrangements like that on the right of figure a diffusor plate in combination with a Fabry-Perot plate can be used as a simple means to check linewidth and mode structure of lasers.

Finesse

It is costumary to define a numerical value which characterizes the width – or better the sharpness – of the maxima. This number is called Finesse of an interferometer and defined as the ratio of peak distance to peak halfwidth.

To get it, the points (phases δ) where the transfer function reaches a value of 0.5 have to be calculated,

$$\frac{1}{1+Q_R\sin^2\frac{\delta}{2}}\overset{!}{=}0.5,$$

$$Q_R\sin^2\frac{\delta}{2}=1 \text{ and for } |\delta-2N\pi|\ll 1: \quad \delta=\pm\frac{2}{\sqrt{Q_R}}.$$

The distance (phase difference) between two peaks is 2π , thus the Finess F is,

$$F=\frac{\pi}{2}\sqrt{Q_R}=\frac{\pi\sqrt{R}}{1-R}.$$

As only the influence of the reflectivity on the linewidth is considered here, often the term Reflectivity Finesse is used to distinguish it from other properties influencing the transfer function.

Contrast

To rate the suppression between maxima, the Contrast is defined as ratio of peak height to the minimum intensity. The transmission minimum $\delta=\pi$ and at equivalent phases define a Contrast value C of,

$$C=Q_R+1.$$

Resolution

The resolution of an optical instrument is defined by the bandwidth of a spectral line, i. e. the broadening imposed on the line by the instrument. We have to calculate what spectral linewidth $\Delta\lambda$ is equivalent to this instrumental linewidth. As hitherto, we assume normal incidence. At the maxima, the phase difference δ is defined,

$$\delta=2kdn=\frac{4\pi}{\lambda}dn=2N\pi.$$

The linewidth can be expressed by the finesse,

$$\Delta\delta=\frac{2\pi}{F},$$

and equated with a spectral linewidth,

$$\Delta\delta = \frac{2\pi}{F} = \frac{4\pi}{\lambda} dn - \frac{4\pi}{\lambda + \Delta\lambda} dn.$$

Thus the resolving power of the instrument is defined by the product of the finesse F and the interference order N,

$$\frac{\lambda}{\Delta\lambda} = \frac{2F}{\lambda} dn = F \cdot N.$$

Equation $\frac{\lambda}{\Delta\lambda} = \frac{2F}{\lambda} dn = F \cdot N.$ shows that, with a finesse of 100, the resolving power of a 100 mm wide grating is exceeded for dn > 0:5 mm. Using air-spaced Fabry-Perot interferometers, values for d up to more than 100 mm can be easily achieved. The resolution then – expressed as frequency – is around 10 MHz.

Free Spectral Range

Taking equation $\frac{\lambda}{\Delta\lambda} = \frac{2F}{\lambda} dn = F \cdot N.$, the resolution of a Fabry-Perot plate can be improved by increasing dn, the optical path difference between the two reflecting surfaces. But, doing this, also the interference order is increased, leading to more problems with overlapping orders. As a measure for the useful working range (no overlapping orders) the Free Spectral Range of an instrument is defined. It is calculated by equating the phase shift for a certain wavelength λ, order N + 1 with that for λ + FSR, order N:

$$\frac{2dn}{\lambda} = \frac{2dn}{\lambda + FSR} + 1$$

$$FSR = \Delta\lambda_{FSR} = \frac{\lambda^2}{2dn}.$$

Compared to the resolution,

$$\frac{\Delta\lambda_{FSR}}{\Delta\lambda_{res}} = F.$$

Tunable Fabry-perot Interferometers

Parallel-plate Fabry-Perot interferometers, often denoted as Etalons, can be tuned by inclining them with respect to the beam direction, thus varying the phase shift inside the plate. This principle is used, e. g., to select single modes inside a laser resonator. For spectroscopic applications, yet, different schemes have to be applied.

The Air-spaced Fabry-perot Interferometer

Nearly all tuning schemes use interferometers where two highly reflective glass plates are separated by a plane-parallel 'plate' of air. The scheme is complementary to the plane-parallel plate discussed and obeys the same equations.

Figure shows the principle, the two plates usually are slightly wedge-shaped to suppress interference due to the outer surfaces. As the optical distance between the two plates is $d \cdot n$, a convenient way to change it is to vary the air pressure. That was classically used, keeping the distance d constant by mechanical spacers and varying the refractive index n of the air by changing the pressure. Nowadays, this technique is – at best – only of historical interest.

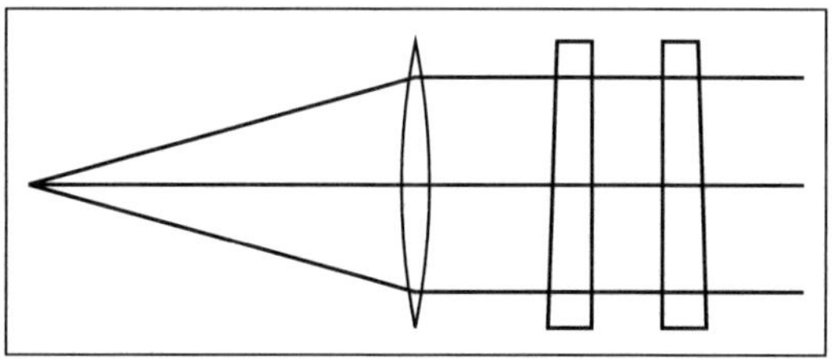

Air-spaced Fabry-Perot interferometer. Parallel light is achieved using a point source in the focus of the illuminating lens.

The Scanning Fabry-perot Interferometer

The preferred method used for scanning is to move one of the two reflecting plates mechanically. The principle is shown in figure As the mechanical scanning distance necessary is in the order of the light wavelength this scanning can be done by piezoelectric actuators.

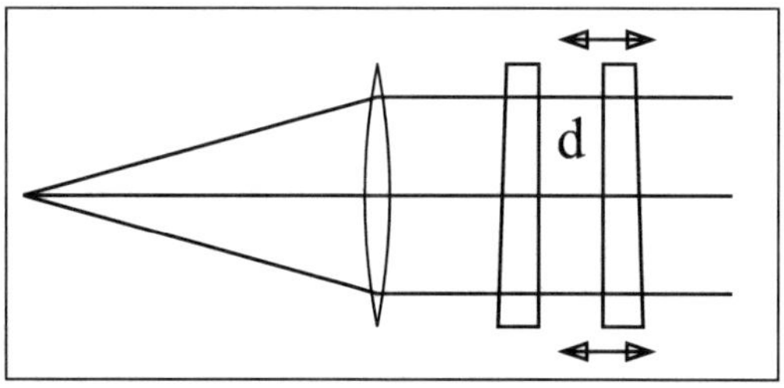

Scanning Fabry-Perot interferometer, principle. The space d between the two reflecting surfaces is varied by moving on of the plates.

The actuators can be driven at moderate voltages (100 . . . 500 V), to get linear scans, usually sawtooth waveforms are used.

Piezoelectric scanning has a variety of advantages compared to other scanning techniques:

- Both, slow and fast scanning modes are possible. In fast scanning modes, the interferogram may be displayed online on an oscilloscope.

- The scanning voltage can be directly used as measure for the plate position.
- The final adjustment can be easily done by automatically or manually adjusting the drive voltages of the (usually three) piezo stacks.
- Stabilization schemes can be easily included by controlling the drive voltages of the piezo stacks. The controller, e. g., can try to hold a constant position and height of the peaks.

There is also one disadvantage of piezoelectric actuators, they often show a large temperature drift, thus they have to be driven in controlled mode.

Pinhole and Divergence

So far we have considered the ideal case of an exact plane wave or exactly parallel light beams. One can construct this ideal case using a point source in the focus of an illuminating lens. In a real-world system, yet, one always has to deal with extended light sources. To approach the ideal case, pinholes are used as an approximation to a point source. The light throughput of the instrument is proportional to the pinhole area, but, increasing the pinhole size, also increases the divergence of the light in the instrument. Thus, to maximize the efficiency, the pinhole size has to be optimized.

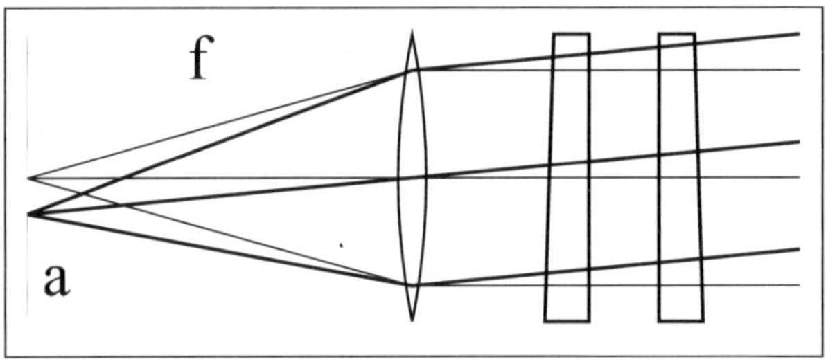

Divergence imposed by a pinhole of radius a.

From figure, the angle between central and outermost beams can be derived to be,

$$\alpha = \arcsin\frac{a}{f} \approx \frac{a}{f}.$$

This results in a corresponding path difference,

$$\Delta d = d\left(\frac{1}{\cos\alpha} - 1\right) \approx d\frac{\alpha^2}{2} \approx d\frac{a^2}{2f^2},$$

which will impose an additional phase shift and thus broaden the peaks in the interferogram. To keep a good resolution, the additional broadening due to this path difference should be less than the halfwidth of the peak,

$$\Delta\delta_{Pinhole} < \Delta\delta_{linewidth},$$

$$\Delta\delta = \frac{2\pi}{F}, \text{ and } \Delta d = d\left(\frac{1}{\cos\alpha} - 1\right) \approx d\frac{\alpha^2}{2} \approx d\frac{a^2}{2f^2},$$

$$2k\Delta d = \frac{2\pi}{\lambda} d \frac{a^2}{f^2} < \frac{2\pi}{F},$$

for the pinhole radius,

$$a < f\sqrt{\frac{\lambda}{Fd}}.$$

For typical geometry data – d = 5 mm, f = 200 mm, λ = 0.5 μm, F = 100 – this results in a radius a < 200 μm.

The influence of the pinhole on the linewidth of a Fabry-Perot interferometer can be denoted as Pinhole Finesse which is defined as

$$F_P = \frac{\lambda}{d}.\frac{f^2}{a^2}.$$

Plate Flatnes

Due to the multiple reflections in a Fabry-Perot interferometer, deviations in the homogeneity of the reflecting surfaces are 'multiplied', too.

Interference Filters

Recent technological achievements in bandpass filter design have led to the relatively inexpensive construction of thin-film interference filters featuring major improvements in wavelength selection and transmission performance. These filters operate by transmitting a selected wavelength region with high efficiency while rejecting, through reflection and destructive interference, all other wavelengths.

Modern interference filters are modeled after the Fabry-Perot interferometer, designed in the late 1800s by Charles Fabry and Alfred Perot, and are constructed with several layers of thin films applied to an optically flat transparent glass surface. The original interferometer consisted of a device having two partially transparent mirrors separated by a small air gap whose size could be varied by translating one or both of the mirrors. Today, more sophisticated interferometers utilize a variety of mechanisms to measure the interference between light beams, and are often employed to monitor thin-film thickness during fabrication of interference filters and mirrors.

Interference filters can be produced with very sharp transmission slopes, which result in steep cut-on and cut-off transition boundaries that greatly exceed those exhibited by standard absorption filters. To produce modern interference filters, successive layers of dielectric materials, with thickness values ranging between one-quarter and one-half of the target wavelength, are deposited onto an optically flat glass or polymer surface in a vacuum. Light that is incident on the multi-layer dielectric surface is either transmitted through the filter with constructive reinforcement, or reflected and reduced in magnitude by destructive interference. The filter bandpass, which is determined by the nature of the layered dielectric surface, determines the wavelengths of light that are allowed to be transmitted and multiply reflected when passing through the filter. Blocked wavelengths that not reinforced and passed by the filter are reflected away and removed from the optical path.

Reflection and Transmission by Interference Filters

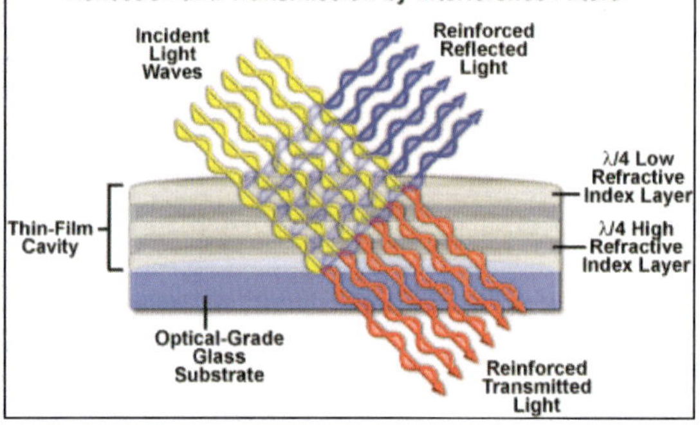

The dielectric materials utilized to fabricate interference filters are generally nonconductive materials having a specific refractive index. Traditional bandpass interference filters are manufactured using zinc sulfide, zinc selenide, or sodium aluminum fluoride (also termed cryolite), but these coatings are hygroscopic and must be insulated from the environment by a protective coating. In addition, the zinc and cryolite salts suffer from low filter transmission characteristics and temperature instability, which further reduces their performance, even though they are simple and relatively cheap to manufacture. After deposition of the thin film salt layers, a final layer of glass or an abrasion-resistant protective coating of silicon monoxide is added.

Introducing semi-transparent layers of metal oxides (known as hard coatings) into thin-film coating technology has alleviated many of the environmental problems associated with interference filters, and dramatically improved their temperature stability. The thin coatings of metals and salts, each with a unique refractive index, are applied in successive layers having alternating high and low refractive index values. The critical element of this design is the interface between two dielectric materials of differing refractive index (one much higher than the other), which is responsible for partially reflecting incident light forward and backward through the filter and producing the interference effect that results in wavelength selection. Reinforced and transmitted wavelength values are determined by the thickness and refractive index of the interspersed dielectric layers. Even though the thin coatings themselves are transparent, light waves reflected and transmitted by the dielectric materials interfere to produce brilliant colors that appear to be emanating from the filter surface.

Dielectric coatings are often bundled into units termed cavities, which are constructed of three to five alternating salt and metal oxide (and sometimes pure metal) layers separated by a wider layer of magnesium fluoride termed a spacer. The spacers are produced with a thickness that corresponds to even multiples of a quarter or half wavelength in order to reflect or transmit light in registration with the dielectric layers. Increasing the number of cavities utilized to build an interference filter produces a proportional increase in the slope of the cut-on and cut-off wavelength transmission boundaries. Filters featuring up to 15 stacked cavities can have a total of over 75 individual dielectric layers and display bandwidths only a few nanometers in size.

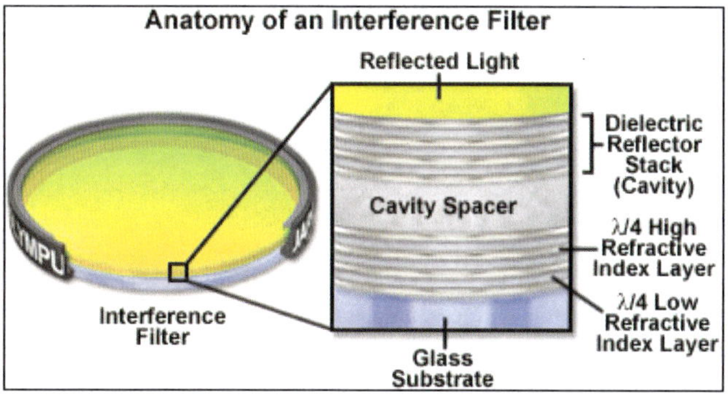

Virtually any type of filter can be designed and constructed with thin-film interference coating technology, including bandpass, shortpass, longpass, dichroic beamsplitters, neutral density, and a variety of mirrors. The number of layers and cavities is utilized to control, with very high precision, the nominal wavelength, bandwidth, and blocking level of the filter. Filters with multiple transmission bands, such as the complex triple-band filters so popular for fluorescence microscopy, can be fabricated with this technique.

The high degree of blocking obtained with thin-film interference filters only applies to a finite wavelength range, beyond which effective blocking drops off dramatically. The range can be extended by adding auxiliary components, such as wideband blockers, but often at a compromise in peak transmission values. In addition, the coating materials utilized in thin-film production are limited in their range of transparency. Once the range has been exceeded, these coatings can become highly absorbing rather than highly reflecting or transmitting, thus reducing the efficiency of the filter. Coating absorption characteristics can also be wavelength-dependent, so the same coating utilized for longpass filters will usually not perform adequately at lower wavelengths in the violet and ultraviolet region. Finally, interference thin-film coatings are sensitive to the incident angle of illumination. As this angle is increased, the spectral characteristics of the coating tend to shift towards shorter wavelengths (the spectrum is blue-shifted). Another drawback is that interference coatings often produce polarized light at high incident angles, an effect that is not always desirable. Regardless of the deficiencies found in thin-film coatings, this technology is still one of the most suitable for wavelength selection in a wide variety of applications.

Coherence

Coherence is one of the most important concepts in optics and is strongly related to the ability of light to exhibit interference effects. A light field is called coherent when there is a fixed phase relationship between the electric field values at different locations or at different times. Partial coherence means that there is some (although not perfect) correlation between phase values. There are various ways of quantifying the degree of coherence.

It is also common to call certain processes or techniques coherent or incoherent. In that case, "coherent" essentially means phase-sensitive. For example, the general method of coherent beam combining relies on the mutual coherence of beams, whereas spectral (incoherent) beam combining does not.

Spatial vs. Temporal Coherence

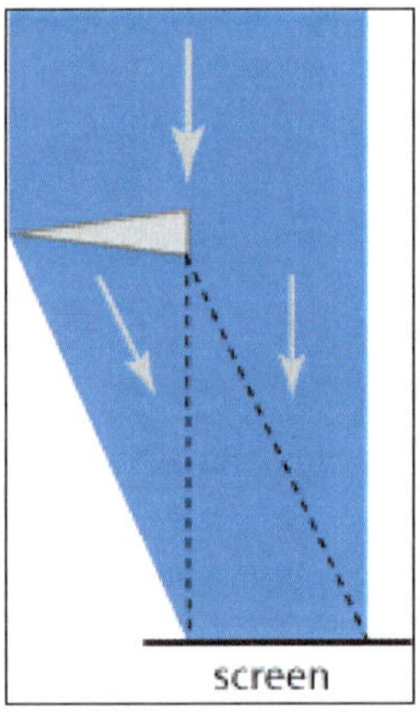

A prism is inserted into a spatially coherent laser beam, generating an interference pattern on the screen.

There are two very different aspects of coherence:

- Spatial coherence means a strong correlation (fixed phase relationship) between the electric fields at different locations across the beam profile. For example, within a cross-section of a beam from a laser with diffraction-limited beam quality, the electric fields at different positions oscillate in a totally correlated way, even if the temporal structure is complicated by a superposition of different frequency components. Spatial coherence is the essential prerequisite of the strong directionality of laser beams.

- Temporal coherence means a strong correlation between the electric fields at one location but different times. For example, the output of a single-frequency laser can exhibit a very high temporal coherence, as the electric field temporally evolves in a highly predictable fashion: it exhibits a clean sinusoidal oscillation over extended periods of time.

Lasers have the potential for generating beams (e.g. Gaussian beams) with very high spatial coherence, and this is perhaps the most fundamental difference between laser light and radiation from other light sources. High spatial coherence arises from the existence of resonator modes, which define spatially correlated field patterns. In situations where only a single resonator mode has sufficient laser gain to oscillate, a single longitudinal mode can be selected, obtaining single-frequency operation with very high temporal coherence as well.

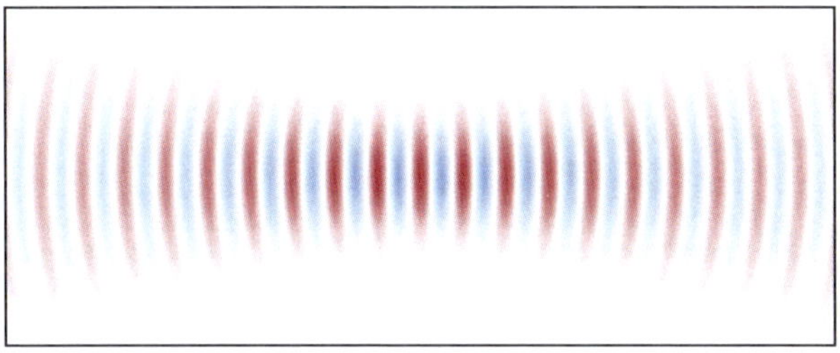

Electric field distribution around the focus of a Gaussian laser beam (coming e.g. from the left side) with perfect spatial and temporal coherence.

Figure shows a beam with high spatial coherence, but poor temporal coherence. The wavefronts are formed as above, and the beam quality is still very high, but the amplitude and phase of the beam varies along the propagation direction. Note that both the local amplitude and the spacing of the wavefronts vary to some extent. Such a beam can be generated e.g. from the output of a supercontinuum source.

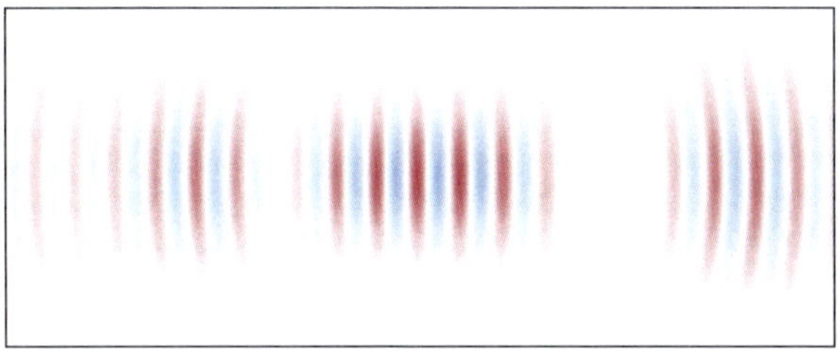

A laser beam with high spatial coherence, but poor temporal coherence.

Figure shows a laser beam with reduced spatial coherence, but high temporal coherence. The wavefronts are deformed, and this results in a high beam divergence and poor beam quality. On the other hand, the beam is monochromatic, so that the spacing of the deformed wavefronts remains constant. Such a beam can result from a single-frequency laser, when its output is sent through some optically inhomogeneous material.

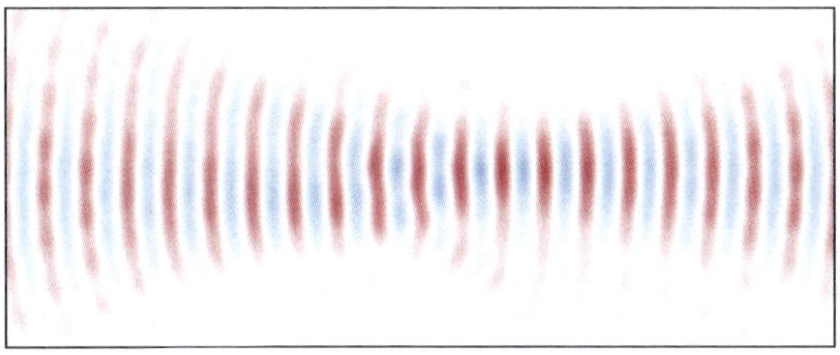

A laser beam with poor spatial coherence, but high temporal coherence.

If a laser beam with high spatial coherence is sent through an optical diffuser element (for example, a very inhomogeneous piece of glass) which totally scrambles the wavefront, the resulting distorted beam could in principle still be considered as spatially coherent, since the phase relationship between the electric fields at different points would still be fixed, as long as the wavefront distortions do not change with time. It would then also in principle be possible to restore a simple beam shape by applying another optical element which compensates the complicated spatial distortions. For most practical purposes, however, such a distorted beam will be spatially incoherent. Nevertheless, a real destruction of spatial coherence requires time-dependent wavefront distortions, as can be obtained with a rotating diffuser, for example. In that case, the temporal coherence is also somewhat degraded.

Quantifying Coherence

There are different ways to quantify the degree of coherence:

- Correlation functions specify the degree of correlation as a function of a spatial or temporal distance. There are correlation functions of different orders. The first-order correlation function is related to the optical spectrum. The second-order correlation function describes intensity correlations, i.e., effects like photon bunching or anti-bunching. Higher-order functions describe still more subtle details.
- The coherence time quantifies the degree of first-order temporal coherence via the time over which coherence is lost.
- The coherence length is the coherence time times the vacuum velocity of light, and thus also characterizes the temporal (not spatial!) coherence via the propagation length (and thus propagation time) over which coherence is lost.
- The linewidth of a single-frequency laser is also strongly related to temporal coherence: a narrow linewidth (high monochromaticity) means high temporal coherence.
- Fringe visibility parameters essentially specify the visibility (contrast) of an interference pattern generated by superposition of two electric fields.

The relationship between optical bandwidth and temporal coherence can be non-trivial. For example, a pulse train from a mode-locked laser can have a broad overall bandwidth, with the Fourier spectrum consisting of discrete very narrow lines (→ *frequency combs*). The temporal coherence can be very high in the sense that there are strong field correlations for large time delays which are close to integer multiples of the pulse period.

Importance of Coherence in Applications

Some applications need light with very high spatial and temporal coherence. This applies, e.g., to many variations of interferometry, holography, and some types of optical sensors (e.g. fiber-optic sensors). Such features are also important for the technique of coherent beam combining.

For other applications, the coherence of the light used should be as low as possible. For example, very low temporal coherence (but combined with high spatial coherence) is required for optical coherence tomography, where images are created with a kind of interferometry, and a high spatial resolution requires low temporal coherence. Suitable light sources for such applications can be based on amplified spontaneous emission (ASE) from a laser amplifier (→ superluminescent sources) or on supercontinuum generation in nonlinear media. A low degree of temporal coherence can also be beneficial for laser projection displays, imaging and pointer applications, as it reduces the tendency for laser speckle and similar interference effects.

Coherence in Quantum Optics

In quantum optics, the term coherence is often used for the state of light-emitting atoms or ions. In that case, coherence refers to a phase relationship between the complex amplitudes corresponding to electronic states. This is important, e.g., in the context of lasing without inversion. There is also the term "coherent states" of the light field, which has yet another meaning.

Fourier Transform Spectroscopy

Fourier transform spectroscopy is a method where one computes an optical spectrum from raw data by applying a Fourier transform algorithm. The method is applied in various techniques for spectroscopy – most often in the context of infrared spectroscopy.

The term time domain spectroscopy is also common, because the measured interference signal is measured in the time domain e.g. in the sense that an optical time delay is varied.

Operation Principle

The operation principle of Fourier transform spectroscopy in its most common form is fairly simple to understand. The investigated electromagnetic radiation (most frequently, infrared light) is sent to an interferometer, normally in the form of a Michelson interferometer. One then measures the optical power at the output of the interferometer as a function of the arm length difference, using some photodetector. That arm length difference is usually manipulated by mechanically moving a mirror (or more conveniently a retroreflector) over some distance.

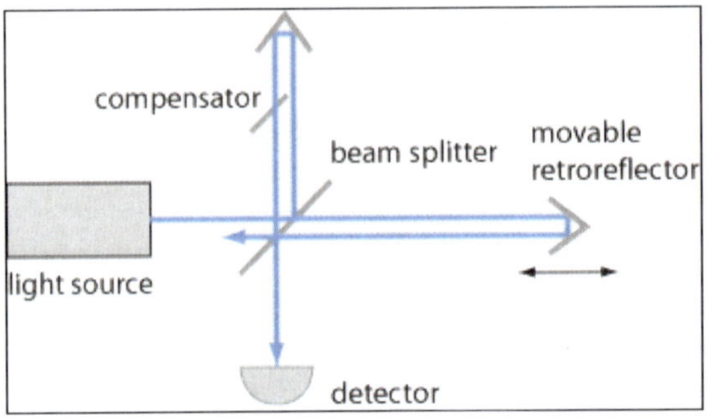

Optical setup of a Fourier transform spectrometer.

If the optical input to the interferometer were monochromatic, one would obtain a sinusoidal oscillation of the detected power as a function of arm length difference, and the period of that oscillation would be the optical wavelength. If the light is polychromatic, the recorded interferogram will be a superposition of contributions from the different wavelength components. Therefore, it is clear that by applying a Fourier transform to those data one can retrieve the optical spectrum – more precisely, the power spectral density as a function of optical frequency or wavelength. Information on the spectral phase is not obtained.

For a mathematically founded understanding, consider that the interferogram signal, resulting from the superposition of two optical electric fields with a certain time delay can be expressed as follows:

$$I(\tau)\left\langle \left|E(t)+E(t-\tau)\right|^{2}\right\rangle$$

Where $I(\tau)$ can be the intensity of the interference signal or alternatively a photocurrent. That signal can be decomposed into a constant and a τ-dependent part; the latter is:

$$I(\tau)=2\left\langle E(t)\cdot E(t-\tau)\right\rangle$$

This is essentially just the autocorrelation of the electric field. According to the Wiener–Kinchine theorem, the Fourier transform of that is the intensity spectrum of the electric field, i.e., the optical spectrum.

The explained operation principle can easily be adapted for absorption spectroscopy. One can record an optical spectrum with and without a specimen inserted into the beam path – before or after the interferometer – and compare the computed spectral intensities to obtain the absorption of the sample in a wide range of wavelengths. More precisely, one obtains the loss of spectral intensity caused by the sample, which may not only be caused by absorption of the specimen but also by surface reflections, for example.

Note that there are also other, less common forms of Fourier transform spectroscopy. For example, terahertz waveforms can be recorded in the time domain with an optical sampling technique based on a photoconductive antenna. One can then apply a Fourier transform to obtain the optical spectrum of a terahertz pulse, in that case also obtaining the spectral phase.

Various Practical Aspects

Required Spatial Range and Resolution

The obtained spectral resolution is limited by the maximum optical path length difference. This is easy to see considering the properties of discrete Fourier transforms, or simply recognizing that the range of path length differences determines the number of oscillation cycles which can be counted. Quantitatively, the resolution in terms of spectroscopic wavenumber is the inverse of the maximum optical path length difference. Simple instruments may work with only a few centimeters of path length difference, achieving spectral resolutions of somewhat better than 1 cm^{-1}, while high precision spectrometers work with much longer path length differences, e.g. several meters.

On the other hand, the maximum wavenumber is half the inverse spatial resolution of the measured path length difference. Therefore, a not particularly high spatial resolution is required for instruments working only with relatively long optical wavelengths, while UV instruments are more demanding in that respect. The spatial *accuracy*, however, should be much higher.

For calculations, note that the variation of pass length difference in a Michelson interferometer is *twice* the amount of movement of a retroreflector.

Light Source

For spectrometers as used in infrared spectroscopy, often uses a very broadband light source for measuring optical properties of samples in a wide wavelength region. The light source should of course have a sufficiently high spectral flux and emit continuously with stable optical properties throughout the interferometer scan. For the near infrared, incandescent lamps are suitable, but there emission is limited to wavelengths below roughly 5 μm by the transmissivity of the bulb glass. For longer wavelength regions, and therefore uses for millimeters not requiring a glass bulb – for example, Nernst glowers based on an electrically heated rod made of zirconium/yttrium ceramics. Silicon carbide rods can even be used up to about 40 μm. Also there are mercury vapor lamps.

Beam Preparation

For the interferometer to work properly, one requires a light beam with high enough spatial coherence. This is because different spatial components of a beam can produce different contributions to the interferogram, effectively washing out the pattern.

Ideally, one would have a Gaussian beam from a laser source. In practice, however, one often deals with incoherent sources, where the light has to be spatially filtered, accepting some loss of optical power. However, the possible power throughput is still substantially better than for a grating monochromator as used in other forms of spectroscopy, where light needs to be fed through a narrow optical slit. This is called the *Jacquinot advantage*, named after Pierre Jacquinot who identified it.

The Beam Splitter

The optical components of the interferometer should of course properly work over the full spectral region of interest. The most substantial challenge arises from the beam splitter, which would ideally exhibit a 50:50 splitting ratio for all relevant wavelengths. That is not strictly required, but it should at least not lead to highly asymmetric splitting or introduce high power losses e.g. by absorption in a substrate. In infrared spectroscopy, one often uses beam splitters with calcium fluoride (CaF_2) substrates for wavelengths up to 8 μm. KBr-based beam splitters with a germanium-based coating can be used up to 25 μm wavelength, but that material is hygroscopic and must therefore be carefully protected against moisture. For the far infrared, one often uses polymer films.

Broadband mirrors are more easily obtained; one often uses metal-coated first surface mirrors, because dielectric mirrors could hardly provide the required reflection bandwidth.

Calibration of Arm Length Variations

The interferometer arm length difference is usually varied with a motorized drive, which can normally not be trusted to provide sufficiently accurate variations of the position. Therefore, one often simultaneously records a second interferogram, using light from a narrow-linewidth laser with sufficiently stable wavelength. One can then computationally correct the data for any deviations of the movement from a perfectly linear movement.

Note that it is not sufficient only to have a positional accuracy which allows one to clearly resolve the oscillations of an interferogram. This is because random position errors also limit the signal-to-noise ratio of the obtained spectra. Therefore, it is essential to realize Fourier transform spectrometers with highly accurate opto-mechanics and an accurate reference interferometer. That also provides a very high wavelength accuracy – better than in dispersive instruments.

Calibration of Spectral Power Density

A simple Fourier transform applied to the raw data will generally not deliver a calibrated optical spectrum, mostly because the responsivity of the used photodetector and the reflectivity of the beam splitter are wavelength-dependent; further influences can come from other optical elements of the setup. Such influences do not matter in absorption spectroscopy, because one only compares spectra with and without an absorbing sample, and the obtained intensity ratios are not affected; one only requires sufficiently strong signals for all relevant wavelengths.

When measuring optical spectra of sources, however, one needs to apply a calibration. It may be done, for example, by comparing with the recorded spectrum of a light source with known spectral shape. In the infrared, one often uses black body radiation for that calibration.

In some cases, one may even calibrate a spectrometer for obtaining absolute values of the power spectral density. This is often not easy, however, for example because of influences of the required spatial filtering of the input beam.

Fourier Transform

Discrete Fourier transforms can quite easily and efficiently be computed, using a Fast Fourier Transform (FFT) algorithm. In the simplest form, such an algorithm works with a number of data points which is a power of 2. Even on a relatively simple microprocessor, the FFT computation usually takes much less time than the acquisition of the raw data.

Interference-based methods of spectroscopy have been used already in the early days of optics, for example by Hippolyte Fizeau, who resolved the doublet of the yellow sodium fluorescence line in the 19th} century (→ Fizeau interferometers). However, computations of optical spectra based on Fast Fourier transform have been implemented only from the middle of the 20th century on, when computers became available; first commercial devices appeared in the 1960s.

Reducing the Sensitivity to Mechanical Noise

For the kind of interferometer as explained above, the sensitivity to mechanical noise (vibrations and shocks or inaccuracies of an optical delay line) is quite high. That sensitivity can be massively reduced by using a common-path interferometer based on birefringence. This can be realized, for example, with a simple optical beam path where two polarization components finally interfere at a polarizer. The optical delay between the two polarization components can be adjusted by moving a wedged birefringent crystal. Because that does not only greatly reduce the sensitivity to vibrations, but also allows very accurate scanning of the delay range, it is particularly suitable for Fourier transform spectroscopy in relatively short wavelength regions.

Applications of Fourier Transform Spectroscopy

The method of Fourier transform spectroscopy is most frequently used in conjunction with infrared light – for the following reasons:

- Particularly in the far infrared, it is difficult to realize focal plane arrays as required for conventional spectrographs, for example. It is thus preferable to use a method where only a simple photodetector is required.

- Due to the limited sensitivity of infrared detectors (particularly at very long wavelengths), it is important to use the light efficiently. It is thus beneficial to avoid excessive power losses at the input slit of a monochromator. Besides, one also enjoys the *Fellgett advantage,* if the measurement noise is dominated by detector noise (e.g. thermal electronic noise) rather than by shot noise, the achievable signal-to-noise ratio is substantially better for the Fourier transform method than for scanning the spectrum with a tunable monochromator, where only a tiny part of the optical spectrum is utilized at any time. This is particularly true in cases where a high spectral resolution is required.
- The Fourier transform method is even somewhat simpler to implement in the infrared, because the required spatial resolution is lower than for visible and ultraviolet light.

The main application of the method is in devices for measuring either optical spectra of light sources or wavelength-dependent properties of materials, such as the transmissivity (e.g. reduced by absorption lines) or the reflectivity.

The principle of Fourier transform spectroscopy is also applied in wavemeters, although those usually deliver only the peak wavelength rather than the full optical spectrum.

There are also applications of the principal in technical fields outside photonics, for example in the context of nuclear magnetic resonance imaging and mass spectroscopy.

References

- Youngs-double-slit-experiment, jee: byjus.com, Retrieved 16 August, 2019
- Multiple-beam-interferometry, microscopy-basics: microscopyu.com, Retrieved 22 May, 2019
- Carey, W. M. (2009). "Lloyd's Mirror—Image Interference Effects". *Acoustics Today*. 5 (2): 14. doi:10.1121/1.3182842
- Interference, filters, java, microscope-resource: olympus-lifescience.com, Retrieved 09 July, 2019
- Fourier-transform-spectroscopy: rp-photonics.com, Retrieved 09 June, 2019

3

Diffraction of Light

Diffraction is defined as the bending of light wave around the edge of an object which depends on the wavelength of light wave. It includes Huygens' principle, Fraunhofer diffraction, single and double slit diffraction, diffraction grating, Fresnel diffraction, etc. The topics elaborated in this chapter will help in gaining a better perspective about these aspects of refraction of light.

Diffraction is the slight bending of light as it passes around the edge of an object. The amount of bending depends on the relative size of the wavelength of light to the size of the opening. If the opening is much larger than the light's wavelength, the bending will be almost unnoticeable. However, if the two are closer in size or equal, the amount of bending is considerable, and easily seen with the naked eye.

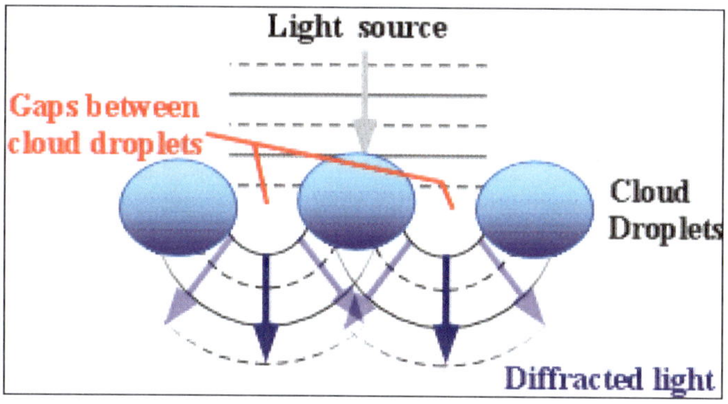

In the atmosphere, diffracted light is actually bent around atmospheric particles most commonly, the atmospheric particles are tiny water droplets found in clouds. Diffracted light can produce fringes of light, dark or colored bands. An optical effect that results from the diffraction of light is the silver lining sometimes found around the edges of clouds or coronas surrounding the sun or moon. The illustration above shows how light (from either the sun or the moon) is bent around small droplets in the cloud.

Optical effects resulting from diffraction are produced through the interference of light waves. To visualize this, imagine light waves as water waves. If water waves were incident upon a float residing on the water surface, the float would bounce up and down in response to the incident waves, producing waves of its own. As these waves spread outward in all directions from the float, they interact with other water waves. If the crests of two waves combine, an amplified wave is produced (constructive interference).

However, if a crest of one wave and a trough of another wave combine, they cancel each other out to produce no vertical displacement (destructive interference).

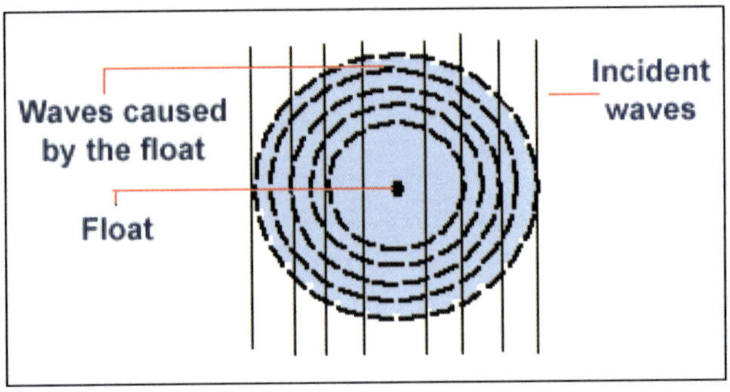

This concept also applies to light waves. When sunlight (or moonlight) encounters a cloud droplet, light waves are altered and interact with one another in a similar manner as the water waves described above. If there is constructive interference, (the crests of two light waves combining), the light will appear brighter. If there is destructive interference, (the trough of one light wave meeting the crest of another), the light will either appear darker or disappear entirely.

Huygens' Principle

A transverse wave, such as an electromagnetic wave like light, as viewed from above and from the side. The direction of propagation is perpendicular to the wavefronts (or wave crests) and is represented by an arrow like a ray.

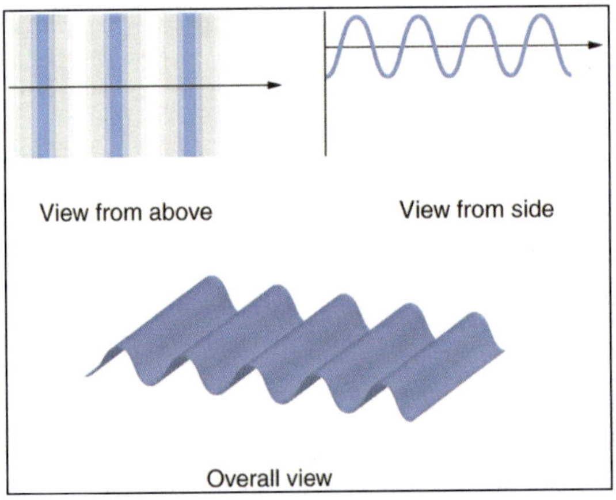

The Dutch scientist Christiaan Huygens developed a useful technique for determining in detail how and where waves propagate. Starting from some known position, Huygens's principle states that:

Every point on a wavefront is a source of wavelets that spread out in the forward direction at the same speed as the wave itself. The new wavefront is a line tangent to all of the wavelets.

Here we shows how Huygens's principle is applied. A wavefront is the long edge that moves, for example, the crest or the trough. Each point on the wavefront emits a semicircular wave that moves at the propagation speed v. These are drawn at a time t later, so that they have moved a distance s = vt. The new wavefront is a line tangent to the wavelets and is where we would expect the wave to be a time later. Huygens's principle works for all types of waves, including water waves, sound waves, and light waves. We will find it useful not only in describing how light waves propagate, but also in explaining the laws of reflection and refraction. In addition, we will see that Huygens's principle tells us how and where light rays interfere.

Huygens's principle applied to a straight wavefront. Each point on the wavefront emits a semicircular wavelet that moves a distance s = vt. The new wavefront is a line tangent to the wavelets.

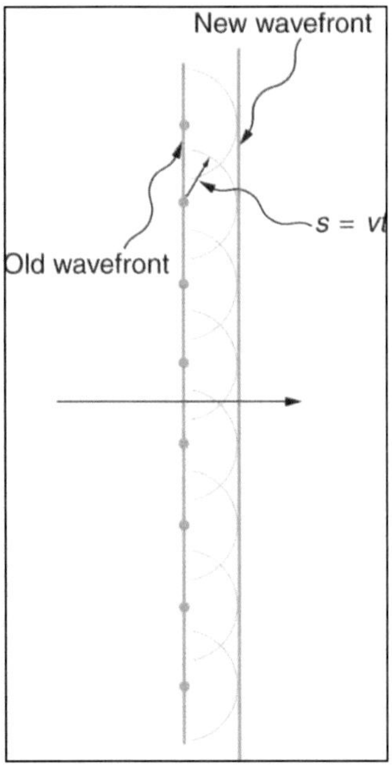

Here we shows how a mirror reflects an incoming wave at an angle equal to the incident angle, verifying the law of reflection. As the wavefront strikes the mirror, wavelets are first emitted from the left part of the mirror and then the right. The wavelets closer to the left have had time to travel farther, producing a wavefront traveling in the direction shown.

Huygens's principle applied to a straight wavefront striking a mirror. The wavelets shown were emitted as each point on the wavefront struck the mirror. The tangent to these wavelets shows that the new wavefront has been reflected at an angle equal to the incident angle. The direction of propagation is perpendicular to the wavefront, as shown by the downward-pointing arrows.

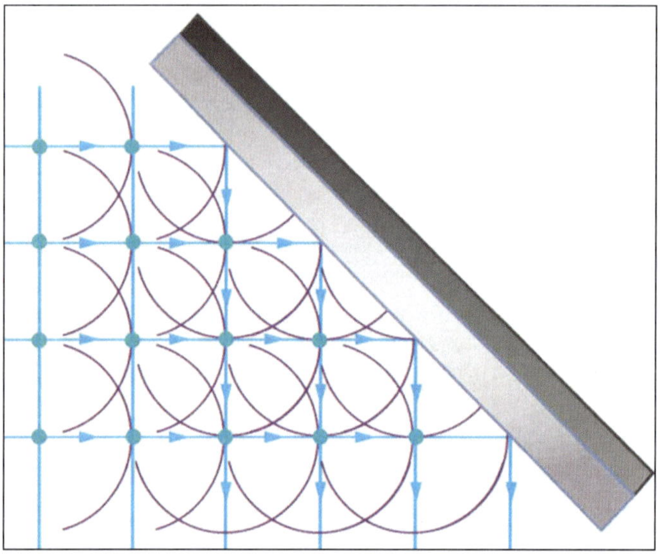

The law of refraction can be explained by applying Huygens's principle to a wavefront passing from one medium to another. Each wavelet in the figure was emitted when the wavefront crossed the interface between the media. Since the speed of light is smaller in the second medium, the waves do not travel as far in a given time, and the new wavefront changes direction as shown. This explains why a ray changes direction to become closer to the perpendicular when light slows down.

Huygens's principle applied to a straight wavefront traveling from one medium to another where its speed is less. The ray bends toward the perpendicular, since the wavelets have a lower speed in the second medium.

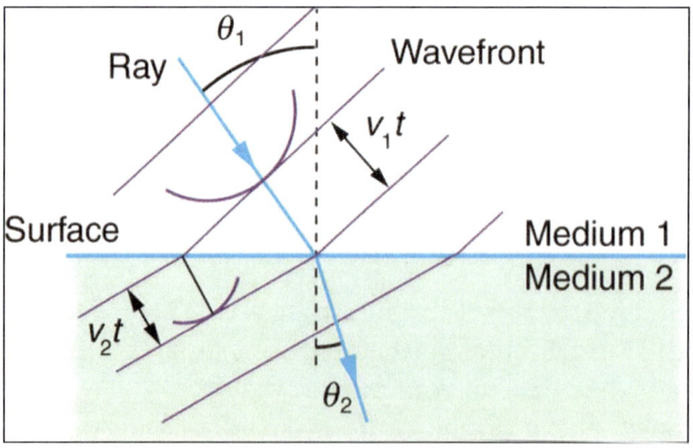

What happens when a wave passes through an opening, such as light shining through an open door into a dark room? For light, we expect to see a sharp shadow of the doorway on the floor of the room, and we expect no light to bend around corners into other parts of the room. When sound passes through a door, we expect to hear it everywhere in the room and, thus, expect that sound spreads out when passing through such an opening. What is the difference between the behavior of sound waves and light waves in this case? The answer is that light has very short wavelengths and acts like a ray. Sound has wavelengths on the order of the size of the door and bends around corners (for frequency of 1000 Hz, $\lambda = c/f = (330m/s)/(1000s^{-1}) = 0.33$ m, about three times smaller than the width of the doorway).

(a) Light passing through a doorway makes a sharp outline on the floor. Since light's wavelength is very small compared with the size of the door, it acts like a ray. (b) Sound waves bend into all parts of the room, a wave effect, because their wavelength is similar to the size of the door.

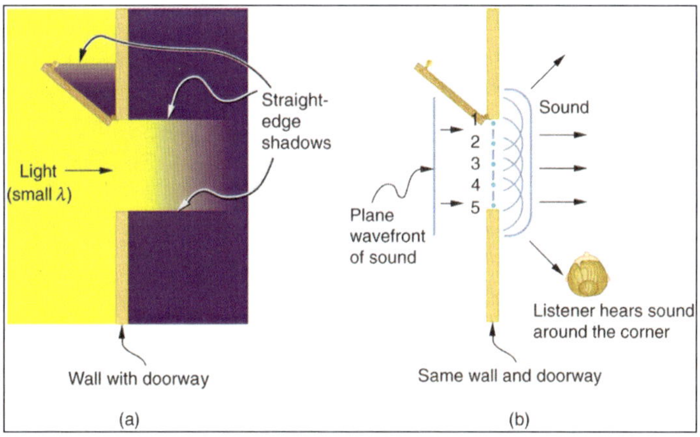

If we pass light through smaller openings, often called slits, we can use Huygens's principle to see that light bends as sound does. The bending of a wave around the edges of an opening or an obstacle is called diffraction. Diffraction is a wave characteristic and occurs for all types of waves. If diffraction is observed for some phenomenon, it is evidence that the phenomenon is a wave.

Huygens's principle applied to a straight wavefront striking an opening. The edges of the wavefront bend after passing through the opening, a process called diffraction. The amount of bending is more extreme for a small opening, consistent with the fact that wave characteristics are most noticeable for interactions with objects about the same size as the wavelength.

Fraunhofer Diffraction

Fraunhofer diffraction is the type of diffraction that occurs in the limit of small Fresnel number $F \ll 1$. In Fraunhofer diffraction, the diffraction pattern is independent of the distance to the screen, depending only on the angles to the screen from the aperture. Let the distance coordinates in the aperture plane be (x', y') and the distance coordinates in the projection plane (x, y). Writing the aperture factor as $\xi(x', y')$. The

Fresnel-Kirchhoff diffraction integral gives the wavefunction $\psi(x, y)$ at the projection plane as,

$$\psi(x,y) = C\int_{\text{aperture}} \xi(x',y')e^{-ik(xx'+yy')/R}dx'dy',$$

Where C is a constant, k is the wavenumber, R is the distance from the aperture to the projection plane, and the integral is taken over the aperture. For $x/R, y/R \ll 1$, can be approximated by the two-dimensional Fourier transform,

$$\psi(u,v) = C\int_{\text{aperture}} \xi(x',y')e^{-ik(ux'+vy')}dx'dy',$$

Where,

$$u \approx \frac{x}{R}$$

$$v \approx \frac{y}{R}$$

are the angular coordinates.

Diffraction by a Circular Aperture

The problem of diffraction at a circular aperture was first solved by Airy in 1835. A circular aperture of diameter 'd ' is shown as AB in figure. A plane wave front WW' is incident normally on this aperture. Every point on the plane wave front in the aperture acts as a source of secondary wavelets. The secondary wavelets spread out in all directions as diffracted rays in the aperture. These diffracted secondary wavelets are converged on the screen SS′ by keeping a convex lens (L) between the aperture and the screen. The screen is at the focal plane of the convex lens. Those diffracted rays traveling normal to the plane of aperture [i.e., along CP_0] are converged at P_0.

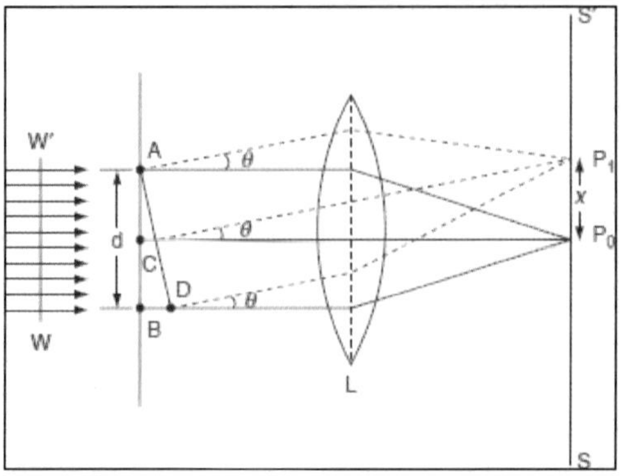

Figure: Fraunhofer diffraction at a circular aperture.

All these waves travel some distance to reach P_0 and there is no path difference between these rays. Hence a bright spot is formed at P_0 known Airy's disc. P_0 corresponds to the central maximum.

Next consider the secondary waves traveling at an angle θ with respect to the direction of CP_0. All these secondary waves travel in the form of a cone and hence, they form a diffracted ring on the screen. The radius of that ring is x and its centre is at P_0. Now consider a point P_1 on the ring, the intensity of light at P_1 depends on the path difference between the waves at A and B to reach P_1. The path difference is BD = AB sin θ = dsin θ. The diffraction due to a circular aperture is similar to the diffraction due to a single slit. Hence, the intensity at P_1 depends on the path difference d sin θ. If the path difference is an integral multiple of λ then intensity at P_1 is minimum. On the other hand, if the path difference is in odd multiples of $\frac{\lambda}{2}$, then the intensity is maximum.

i.e., $d\sin\theta = n\lambda$ for minima,

and $d \sin\theta = (2n-1)\frac{\lambda}{2}$ for maxima,

where n = 1, 2, 3, ... etc. n = 0 corresponds to central maximum.

The Airy disc is surrounded by alternate bright and dark concentric rings, called the Airy's rings. The intensity of the dark ring is zero and the intensity of the bright ring decreases as we go radially from P_0 on the screen. If the collecting lens(L) is very near to the circular aperture or the screen is at a large distance from the lens, then,

$$\sin\theta \approx \theta \approx \frac{x}{f}$$

Where f is the focal length of the lens.

Also from the condition for first secondary minimum [using equation $d\sin\theta = n\lambda$],

$$\sin\theta \approx \theta \approx \frac{\lambda}{d}$$

Equations $\sin\theta \approx \theta \approx \frac{x}{f}$ and $\sin\theta \approx \theta \approx \frac{\lambda}{d}$ are equal,

$$\text{So, } \frac{x}{f} = \frac{\lambda}{d} \text{ or } x = \frac{f\lambda}{d}$$

But according to Airy, the exact value of x is,

$$x = \frac{1.22 f\lambda}{d}$$

using equation $x = \frac{1.22 f\lambda}{d}$ the radius of Airy's disc can be obtained. Also from equation $x = \frac{1.22 f\lambda}{d}$ we know that the radius of Airy's disc is inversely proportional to the diameter of the aperture. Hence by decreasing the diameter of aperture, the size of Airy's disc increases.

Limit of Resolution

Light diffracts as it moves through space, bending around obstacles, interfering constructively and destructively. While this can be used as a spectroscopic tool—a diffraction grating disperses light according to wavelength, for example, and is used to produce spectra—diffraction also limits the detail we can obtain in images. Figure (a) shows the effect of passing light through a small circular aperture. Instead of a bright spot with sharp edges, a spot with a fuzzy edge surrounded by circles of light is obtained. This pattern is caused by diffraction similar to that produced by a single slit. Light from different parts of the circular aperture interferes constructively and destructively. The effect is most noticeable when the aperture is small, but the effect is there for large apertures, too.

(a) Monochromatic light passed through a small circular aperture produces this diffraction pattern. (b) Two point light sources that are close to one another produce overlapping images because of diffraction. (c) If they are closer together, they cannot be resolved or distinguished.

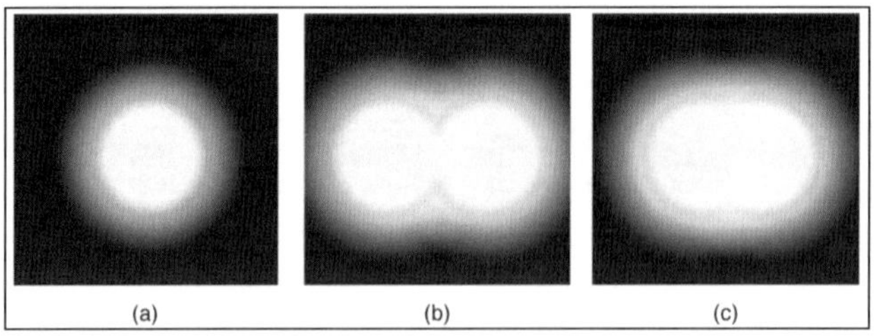

How does diffraction affect the detail that can be observed when light passes through an aperture? Figure (b) shows the diffraction pattern produced by two point light sources that are close to one another. The pattern is similar to that for a single point source, and it is just barely possible to tell that there are two light sources rather than one. If they were closer together, as in figure (c), we could not distinguish them, thus limiting the detail or resolution we can obtain. This limit is an inescapable consequence of the wave nature of light.

There are many situations in which diffraction limits the resolution. The acuity of our vision is limited because light passes through the pupil, the circular aperture of our eye.

Be aware that the diffraction-like spreading of light is due to the limited diameter of a light beam, not the interaction with an aperture. Thus light passing through a lens with a diameter D shows this effect and spreads, blurring the image, just as light passing through an aperture of diameter D does. So diffraction limits the resolution of any system having a lens or mirror. Telescopes are also limited by diffraction, because of the finite diameter D of their primary mirror.

Just what is the limit? To answer that question, consider the diffraction pattern for a circular aperture, which has a central maximum that is wider and brighter than the maxima surrounding it (similar to a slit). It can be shown that, for a circular aperture of diameter D, the first minimum in the diffraction pattern occurs at $\theta = 1.22\lambda / D$ (providing the aperture is large compared with the wavelength of light, which is the case for most optical instruments). The accepted criterion for determining the diffraction limit to resolution based on this angle was developed by Lord Rayleigh in the 19th century. The Rayleigh criterion for the diffraction limit to resolution states that two images are just resolvable when the center of the diffraction pattern of one is directly over the first minimum of the diffraction pattern of the other. The first minimum is at an angle of $\theta = 1.22\lambda / D$, so that two point objects are just resolvable if they are separated by the angle,

$$\theta = 1.22\frac{\lambda}{D},$$

where λ is the wavelength of light (or other electromagnetic radiation) and D is the diameter of the aperture, lens, mirror, etc., with which the two objects are observed. In this expression, θ has units of radians.

(a) Graph of intensity of the diffraction pattern for a circular aperture. Note that, similar to a single slit, the central maximum is wider and brighter than those to the sides. (b) Two point objects produce overlapping diffraction patterns. Shown here is the Rayleigh criterion for being just resolvable. The central maximum of one pattern lies on the first minimum of the other.

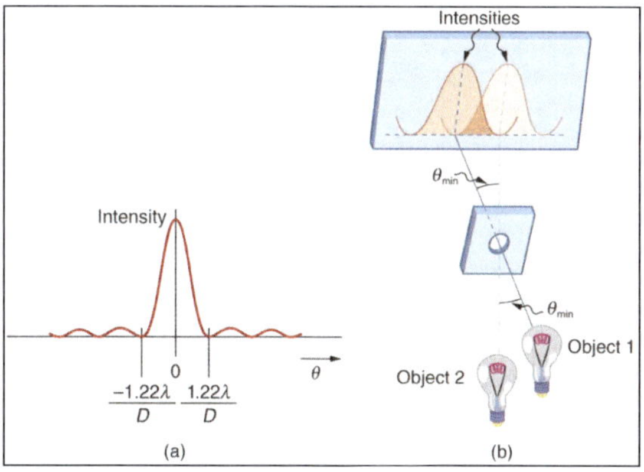

Calculating diffraction limits of the hubble space telescope.

The primary mirror of the orbiting Hubble Space Telescope has a diameter of 2.40 m. Being in orbit, this telescope avoids the degrading effects of atmospheric distortion on its resolution. (a) What is the angle between two just-resolvable point light sources (perhaps two stars)? Assume an average light wavelength of 550 nm. (b) If these two stars are at the 2 million light year distance of the Andromeda galaxy, how close together can they be and still be resolved? (A light year, or ly, is the distance light travels in 1 year.)

Strategy

The Rayleigh criterion stated in the equation $\theta = 1.22\frac{\lambda}{D}$ gives the smallest possible angle θ between point sources, or the best obtainable resolution. Once this angle is found, the distance between stars can be calculated, since we are given how far away they are.

Solution for (a)

The Rayleigh criterion for the minimum resolvable angle is,

$$\theta = 1.22\frac{\lambda}{D}.$$

Entering known values gives,

$$\theta = 1.22\frac{55010^{-9}m}{2.40\text{ m}}$$
$$= 2.8010^{-7}\text{rad}.$$

Solution for (b)

The distance s between two objects a distance r away and separated by an angle θ is $s = r\theta$.

Substituting known values gives,

$$s = \left(2.010^{6}\text{ly}\right)\left(2.8010^{-7}\text{rad}\right)$$
$$= 0.56\text{ ly}$$

Discussion

The angle found in part (a) is extraordinarily small (less than 1/50,000 of a degree), because the primary mirror is so large compared with the wavelength of light. As noticed, diffraction effects are most noticeable when light interacts with objects having sizes on the order of the wavelength of light. However, the effect is still there, and there is a diffraction limit to what is observable. The actual resolution of the Hubble

Telescope is not quite as good as that found here. As with all instruments, there are other effects, such as non-uniformities in mirrors or aberrations in lenses that further limit resolution.

(a) (b)

These two photographs of the M82 galaxy give an idea of the observable detail using the Hubble Space Telescope compared with that using a ground-based telescope. (a) On the left is a ground-based image. (b) The photo on the right was captured by Hubble.

The answer in part (b) indicates that two stars separated by about half a light year can be resolved. The average distance between stars in a galaxy is on the order of 5 light years in the outer parts and about 1 light year near the galactic center. Therefore, the Hubble can resolve most of the individual stars in Andromeda galaxy, even though it lies at such a huge distance that its light takes 2 million years for its light to reach us.

A 305-m-diameter natural bowl at Arecibo in Puerto Rico is lined with reflective material, making it into a radio telescope. It is the largest curved focusing dish in the world. Although D for Arecibo is much larger than for the Hubble Telescope, it detects much longer wavelength radiation and its diffraction limit is significantly poorer than Hubble's. Arecibo is still very useful, because important information is carried by radio waves that is not carried by visible light.

Diffraction is not only a problem for optical instruments but also for the electromagnetic radiation itself. Any beam of light having a finite diameter D and a wavelength λ exhibits diffraction spreading. The beam spreads out with an angle θ given by the equation $\theta = 1.22\frac{\lambda}{D}$. Take, for example, a laser beam made of rays as parallel as possible (angles between rays as close to $\theta = 0$ as possible) instead spreads out at an angle $\theta = 1.22\lambda / D$, where D is the diameter of the beam and λ is its wavelength. This spreading is impossible to observe for a flashlight, because its beam is not very parallel to start with. However, for long-distance transmission of laser beams or microwave signals, diffraction spreading can be significant. To avoid this, we can increase D. This is done for laser light sent to the Moon to measure its distance from the Earth. The laser beam is expanded through a telescope to make D much larger and θ smaller.

The beam produced by this microwave transmission antenna will spread out at a minimum angle $\theta = 1.22\lambda / D$ due to diffraction. It is impossible to produce a near-parallel beam, because the beam has a limited diameter.

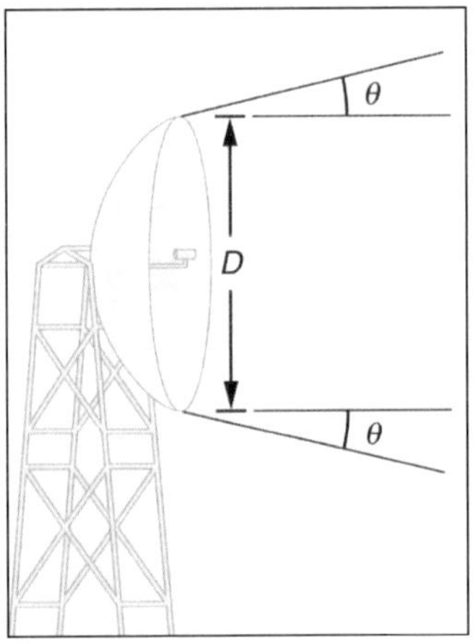

In most biology laboratories, resolution is presented when the use of the microscope is introduced. The ability of a lens to produce sharp images of two closely spaced point objects is called resolution. The smaller the distance x by which two objects can be separated and still be seen as distinct, the greater the resolution. The resolving power of a lens is defined as that distance x. An expression for resolving power is obtained from the Rayleigh criterion. In figure we have two point objects separated by a distance x. According to the Rayleigh criterion, resolution is possible when the minimum angular separation is,

$$\theta = 1.22\frac{\lambda}{D} = \frac{x}{d},$$

Where d is the distance between the specimen and the objective lens, and we have used the small angle approximation (i.e., we have assumed that x is much smaller than d), so that $\tan\theta \approx \sin\theta \approx \theta$.

Therefore, the resolving power is:

$$x = 1.22\frac{\lambda \mathrm{d}}{D}.$$

Another way to look at this is by examining the concept of Numerical Aperture(NA). There, NA is a measure of the maximum acceptance angle at which the fiber will take light and still contain it within the fiber. Figure shows a lens and an object at point P. The NA here is a measure of the ability of the lens to gather light and resolve fine detail. The angle subtended by the lens at its focus is defined to be $\theta = 2\alpha$. From the figure and again using the small angle approximation, we can write,

$$\sin\alpha = \frac{D/2}{d} = \frac{D}{2d}.$$

The NA for a lens is NA = $n\sin\alpha$, where n is the index of refraction of the medium between the objective lens and the object at point P.

From this definition for NA, we can see that,

$$x = 1.22\frac{\lambda \mathrm{d}}{D} = 1.22\frac{\lambda}{2\sin\alpha} = 0.61\frac{\lambda \mathrm{n}}{\mathrm{NA}}.$$

In a microscope, NA is important because it relates to the resolving power of a lens. A lens with a large NA will be able to resolve finer details. Lenses with larger NA will also be able to collect more light and so give a brighter image. Another way to describe this situation is that the larger the NA, the larger the cone of light that can be brought into the lens, and so more of the diffraction modes will be collected. Thus the microscope has more information to form a clear image, and so its resolving power will be higher.

(a) Two points separated by at distance x and a positioned a distance d away from the objective. (b) Terms and symbols used in discussion of resolving power for a lens and an object at point P.

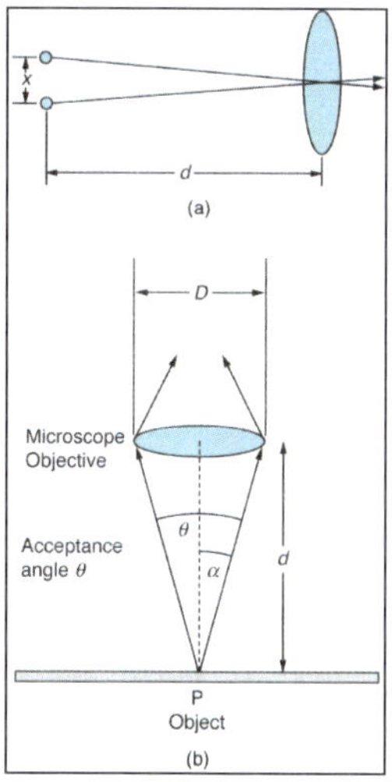

One of the consequences of diffraction is that the focal point of a beam has a finite width and intensity distribution. Consider focusing when only considering geometric optics. The focal point is infinitely small with a huge intensity and the capacity to incinerate most samples irrespective of the NA of the objective lens. For wave optics, due to diffraction, the focal point spreads to become a focal spot with the size of the spot decreasing with increasing NA. Consequently, the intensity in the focal spot increases with increasing NA. The higher the NA, the greater the chances of photode grading the specimen. However, the spot never becomes a true point.

(a) In geometric optics, the focus is a point, but it is not physically possible to produce such a point because it implies infinite intensity. (b) In wave optics, the focus is an extended region.

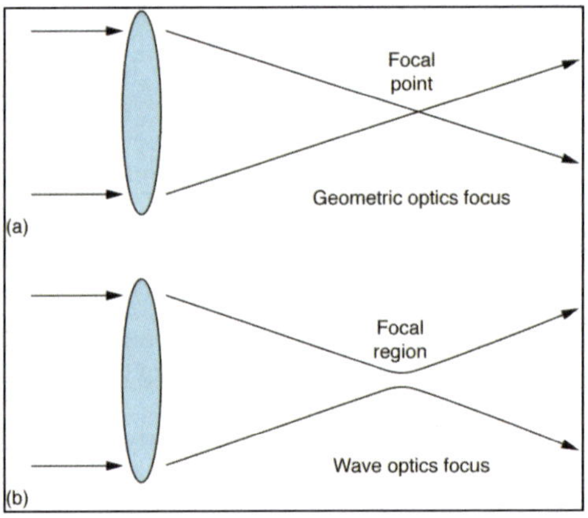

Single-slit Diffraction

Light passing through a single slit forms a diffraction pattern somewhat different from those formed by double slits or diffraction gratings. Figure shows a single-slit diffraction pattern. Note that the central maximum is larger than maxima on either side and that the intensity decreases rapidly on either side. In contrast, a diffraction grating produces evenly spaced lines that dim slowly on either side of the center.

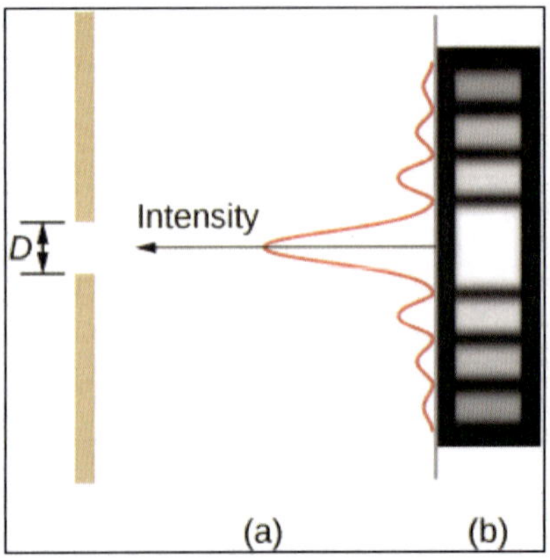

Single-slit diffraction pattern. (a) Monochromatic light passing through a single slit has a central maximum and many smaller and dimmer maxima on either side. The central maximum is six times higher than shown. (b) The diagram shows the bright central maximum, and the dimmer and thinner maxima on either side.

The analysis of single-slit diffraction is illustrated in figure. Here, the light arrives at the slit, illuminating it uniformly and is in phase across its width. We then consider light propagating onwards from different parts of the same slit. According to Huygens's principle, every part of the wave front in the slit emits wavelets. These are like rays that start out in phase and head in all directions. (Each ray is perpendicular to the wave front of a wavelet.) Assuming the screen is very far away compared with the size of the slit, rays heading toward a common destination are nearly parallel. When they travel straight ahead, as in part (a) of the figure, they remain in phase, and we observe a central maximum. However, when rays travel at an angle θ relative to the original direction of the beam, each ray travels a different distance to a common location, and they can arrive in or out of phase. In part (b), the ray from the bottom travels a distance of one wavelength λ farther than the ray from the top. Thus, a ray from the center travels a distance $\lambda/2$ less than the one at the bottom edge of the slit, arrives out of phase, and interferes destructively. A ray from slightly above the center and one from slightly above the bottom also cancel one another. In fact, each ray from the slit interferes destructively with another ray. In other words, a pair-wise cancellation of all rays results in a dark minimum in intensity at this angle. By symmetry, another minimum occurs at the same angle to the right of the incident direction (toward the bottom of the figure) of the light.

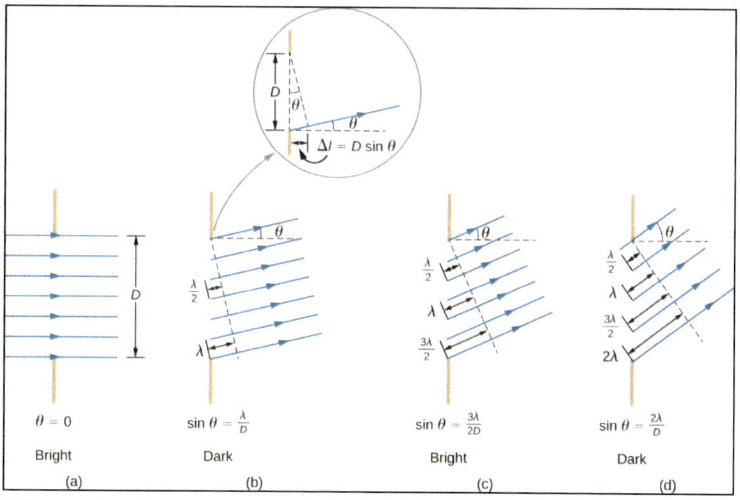

Figure: Light passing through a single slit is diffracted in all directions and may interfere constructively or destructively, depending on the angle. The difference in path length for rays from either side of the slit is seen to be D sin θ.

At the larger angle shown in part (c), the path lengths differ by $3\lambda/2$ for rays from the top and bottom of the slit. One ray travels a distance λ different from the ray from the bottom and arrives in phase, interfering constructively. Two rays, each from slightly above those two, also add constructively. Most rays from the slit have another ray to interfere with constructively, and a maximum in intensity occurs at this angle. However, not all rays interfere constructively for this situation, so the maximum is not as intense as the central maximum. Finally, in part (d), the angle shown is large enough to produce a second minimum. As seen in the figure, the difference in path length for rays from either side of the slit is D sin θ, and we see that a destructive minimum is obtained when this distance is an integral multiple of the wavelength.

Thus, to obtain destructive interference for a single slit,

$$\underbrace{D\sin\theta = m\lambda}_{\text{destructive interference}}$$

Where,

- m = ± 1, ± 2, ± 3, ...,
- D is the slit width,
- λ is the light's wavelength,
- θ is the angle relative to the original direction of the light, and
- m is the order of the minimum.

Figure shows a graph of intensity for single-slit interference, and it is apparent that the maxima on either side of the central maximum are much less intense and not as wide.

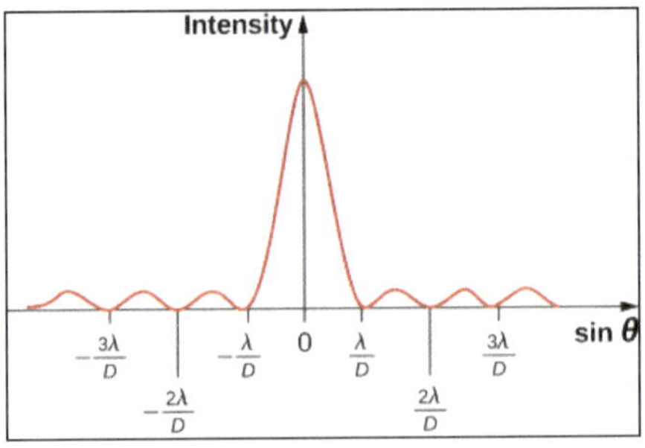

Figure: A graph of single-slit diffraction intensity showing the central maximum to be wider and much more intense than those to the sides. In fact, the central maximum is six times higher than shown here.

Calculating Single-slit Diffraction

Visible light of wavelength 550 nm falls on a single slit and produces its second diffraction minimum at an angle of 45.0° relative to the incident direction of the light, as in figure.

- What is the width of the slit?
- At what angle is the first minimum produced?

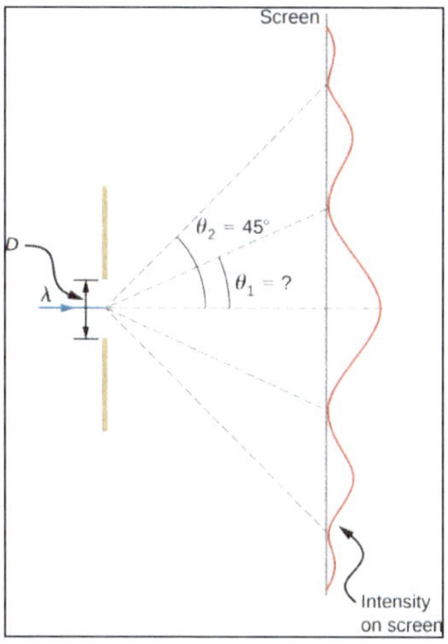

In this example, we analyze a graph of the single-slit diffraction pattern.

Strategy

From the given information, and assuming the screen is far away from the slit, we can use the equation $D \sin \theta = m\lambda$ first to find D, and again to find the angle for the first minimum θ_1.

We are given that $\lambda = 550$ nm, m = 2, and $\theta_2 = 45.0°$. Solving the equation $D \sin \theta = m\lambda$ for D and substituting known values gives,

$$D = \frac{m\lambda}{\sin \theta_2} = \frac{2(550nm)}{\sin 45.0°} = \frac{1100 \times 10^{-9} m}{0.707} = 1.56 \times 10^{-6} m.$$

Solving the equation $D \sin \theta = m\lambda$ for $\sin \theta_1$ and substituting the known values gives,

$$\sin\theta_1 = \frac{m\lambda}{D} = \frac{1(550 \times 10^{-9} m)}{1.56 \times 10^{-6} m}.$$

Thus the angle θ_1 is,

$$\theta_1 = \sin^{-1} 0.354 = 20.7°.$$

Significance

We see that the slit is narrow (it is only a few times greater than the wavelength of light). This is consistent with the fact that light must interact with an object comparable in size to its wavelength in order to exhibit significant wave effects such as this single-slit diffraction pattern. We also see that the central maximum extends 20.7° on either side of the original beam, for a width of about 41°. The angle between the first and second minima is only about 24°(45.0°–20.7°). Thus, the second maximum is only about half as wide as the central maximum.

Double-slit Diffraction

To calculate the diffraction pattern for two (or any number of) slits, we need to generalize the method we just used for a single slit. That is, across each slit, we place a uniform distribution of point sources that radiate Huygens wavelets, and then we sum the wavelets from all the slits. This gives the intensity at any point on the screen.

The double slits have been represented as A_1B_1 and A_2B_2 in figure. The slits are narrow and rectangular in shape. The plane of the slits are perpendicular to plane of the paper. Let the width of both the slits be equal and it is 'e' and they are separated by opaque length 'd'. A monochromatic plane wave front of wave length 'λ' is incident normally on both the slits.

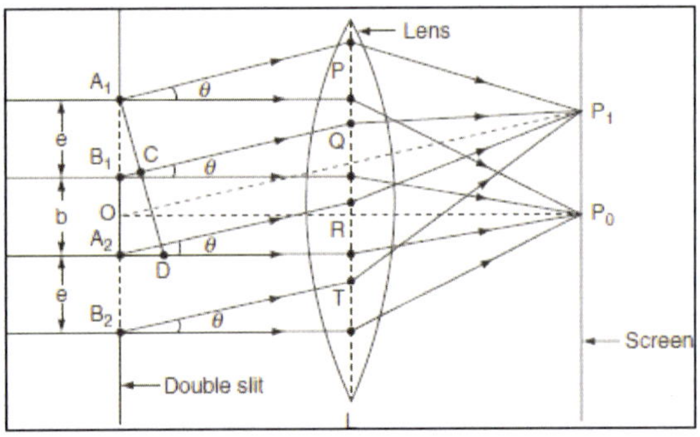

Light is made incident on arrangement of double slit. The secondary wavelets travelling in the direction of OP_0 are brought to focus at P_0 on the screen SS′ by using a converging lens L. P_0 corresponds to the position of the central bright maximum. The intensity distribution on the screen is the combined effect of interference of diffracted secondary waves from the slits.

The diffracted intensity on the screen is very large along the direction of incident beam [i.e along OP_0]. Hence it is maximum at P_0. This is known as principal maximum of zero order.

The intensity at point P_1 on the screen is obtained by applying the Fraunhofer diffraction theory at single slit and interference of diffracted waves from the two slits. The diffracted wave amplitude due to single slit at an angle θ with respect to incident beam is $A\dfrac{\sin\alpha}{\alpha}$, where 2α is the phase difference between the secondary wavelets arising at the end points of a slit. This phase difference can be estimated as follows: Draw a normal from A_1 to B_1Q. Now, B_1C is the path difference between the diffracted waves at an angle 'θ' at the slit A_1B_1.

From the triangle A_1B_1C,

$$\sin\theta = \frac{B_1C}{A_1B_1} = \frac{B_1C}{e} \text{ or } B_1C = e\sin\theta.$$

The corresponding phase difference $(2\alpha) = \frac{2\pi}{\lambda} e\sin\theta$,

$$\text{Or } \alpha = \frac{\pi}{\lambda} e\sin\theta$$

The diffracted wave amplitudes, $\left[A\frac{\sin\alpha}{\alpha}\right]$ at the two slits combine to produce interference. The path difference between the rays coming from corresponding points in the slits A_1B_1 and A_2B_2 can be found by drawing a normal from A_1 to A_2R. A_2D is the path difference between the waves from corresponding points of the slits.

In the triangle $A_1A_2D \frac{A_2D}{A_1A_2} = \sin\theta$ or the path difference $A_2D = A_1A_2 \sin\theta = (e + b)\sin\theta$

The corresponding phase difference $(2\beta) = \frac{2\pi}{\lambda}(e+b)\sin\theta$.

Applying the theory of interference on the wave amplitudes $\left[A\frac{\sin\alpha}{\alpha}\right]$ at the two slits gives the resultant wave amplitude (R).

$$R = 2A\frac{\sin\alpha}{\alpha}\cos\beta$$

The intensity at P_1 is,

$$I = R^2 = 4A^2\frac{\sin^2\alpha}{\alpha^2}\cos^2\beta$$

$$= 4I_\circ\frac{\sin^2\alpha}{\alpha^2}\cos^2\beta$$

$$\left[\text{since } I_\circ = A^2\right]$$

Equation $I = 4I_\circ\frac{\sin^2\alpha}{\alpha^2}\cos^2\beta$ represents the intensity distribution on the screen. The intensity at any point on the screen depends on α and β. The intensity of central maximum is $4I_0$. The intensity distribution at different points on the screen can be explained in terms of path difference between the incident and diffracted rays as follows. In equation $I = 4I_\circ\frac{\sin^2\alpha}{\alpha^2}\cos^2\beta$ the term $\cos^2\beta$ corresponds to interference and $\frac{\sin^2\alpha}{\alpha}$ corresponds to diffraction. Now, we will shall look at the conditions for interference and diffraction maxima and minima.

Interference maxima and minima: If the path difference $A_2D=(e+b)\sin\theta_n=\pm n\lambda$ where n = 1, 2, 3... then 'θ_n' gives the directions of the maxima due to interference of light waves coming from the two slits. The ± sign indicates maxima on both sides with respect to the central maximum. On the other hand if the path difference is odd multiples of $\frac{\lambda}{2}$ i.e., $A_2D=(e+b)\sin\theta_n=\pm(2n-1)\frac{\lambda}{2}$, then θ_n gives the directions of minima due to interference of the secondary waves from the two slits on both sides with respect to central maximum.

Diffraction maxima and minima: If the path difference $B_1C=e\sin\theta n=\pm n\lambda$, where n = 1, 2, 3... then θ_n gives the directions of diffraction minima. The ± sign indicates minima on both sides with respect to central maximum. For diffraction maxima $e\sin\theta_n=\pm(2n-1)\frac{\lambda}{2}$ is the condition. The ± sign indicates maxima on both sides with respect to central maximum.

The intensity distribution on the screen due to double slit diffraction is shown in figure (a) represents the graph for interference term, figure (b) shows the graph for diffraction term and figure (c) represents the resultant distribution.

Based on the relative values of e and b certain orders of interference maxima are missing in the resultant pattern.

The direction of interference maxima are given as $(e+b)\sin\theta_n=n\lambda$ where n = 1, 2, 3, ... and the directions of diffraction minima are given as $e\sin\theta_m=m\lambda$ where m = 1, 2, 3, ...

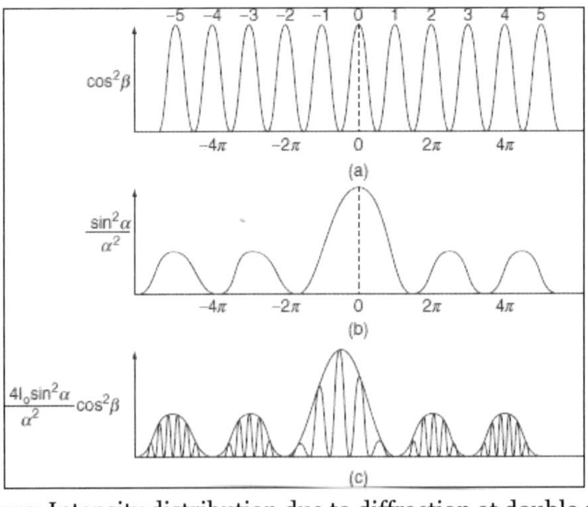

Figure: Intensity distribution due to diffraction at double slit.

For some values of θ_n, the values of e and b are satisfied such that at these positions the interference maxima and the diffraction minima are formed. The combined effect results in missing of certain orders of interference maxima. Now we see certain values of e and b for which interference maxima are missing.

Let e = b

Then, $2\ e\ \sin\theta_n = n\lambda$ and $e\ \sin\theta_m = m\lambda$,

$$\therefore \frac{n}{m} = 2 \text{ or } n = 2m$$

If m = 1, 2, 3 ... then n = 2, 4, 6... i.e., the interference orders 2, 4, 6 ... missed in the diffraction pattern.

If 2e = b

Then $3e\ \sin\theta_m = n\lambda$ and $e\ \sin\theta_m = m\lambda$,

$$\therefore \frac{n}{m} = 3 \quad \text{or} \quad n = 3m$$

if μ = 1, 2, 3... Then n = 3, 6, 9... i.e the interference orders 3, 6, 9... are missed in the diffraction pattern.

if e + b = e,

i.e b = 0

the two slits are joined. So, the diffraction pattern is due to a single slit of width 2e.

Intensity of the Fringes

Figure shows that the intensity of the fringe for m = 3 is zero, but what about the other fringes? Calculate the intensity for the fringe at m = 1 relative to I_0, the intensity of the central peak.

Strategy

Determine the angle for the double-slit interference fringe, using the equation from Interference, then determine the relative intensity in that direction due to diffraction.

We know that the bright interference fringes occur at $d \sin\theta = m\lambda$, or

$$\sin\theta = \frac{m\lambda}{d}.$$

$$I = I_0 \left(\frac{\sin\beta}{\beta}\right)^2, where\ \beta = \frac{phi}{2} = \frac{\pi D \sin\theta}{\lambda}.$$

Substituting from above,

$$\beta = \frac{\pi D \sin\theta}{\lambda} = \frac{\pi D}{\lambda} \cdot \frac{m\lambda}{d} = \frac{m\pi D}{d}.$$

For D = 2λ, d = 6λ, and m = 1,

$$\beta = \frac{(1)\pi(2\lambda)}{(6\lambda)} = \frac{\pi}{3}.$$

Then, the intensity is,

$$I = I_0\left(\frac{\sin\beta}{\beta}\right)^2 = I_0\left(\frac{\sin(\pi/3)}{\pi/3}\right)^2 = 0.684 I_0.$$

Significance

Note that this approach is relatively straightforward and gives a result that is almost exactly the same as the more complicated analysis using phasors to work out the intensity values of the double-slit interference. The phasor approach accounts for the downward slope in the diffraction intensity so that the peak near m=1 occurs at a value of θ ever so slightly smaller than we have shown here.

Two-slit Diffraction

Suppose that in Young's experiment, slits of width 0.020 mm are separated by 0.20 mm. If the slits are illuminated by monochromatic light of wavelength 500 nm, how many bright fringes are observed in the central peak of the diffraction pattern?

From, the angular position of the first diffraction minimum is,

$$\theta \approx \sin\theta = \frac{\lambda}{D} = \frac{5.0\times10^{-7} m}{2.0\times10^{-5} m} = 2.5\times10^{-2} rad.$$

Using $\sin\theta = m\lambda$ for $\theta = 2.5\times10^{-2} rad$, we find

$$m = \frac{d\sin\theta}{\lambda} = \frac{(0.20mm)(2.5\times10^{-2} rad)}{(5.0\times10^{-7} m)} = 10,$$

which is the maximum interference order that fits inside the central peak. We note that m = ± 10 are missing orders as θ matches exactly. Accordingly, we observe bright fringes for,

m = −9, −8, −7, −6, −5, −4, −3, −2, −1, 0, +1, +2, +3, +4, +5, +6, +7, +8, and +9

for a total of 19 bright fringes.

N-slit Diffraction

An arrangement consisting of large number of parallel slits of the same width and separated by equal opaque spaces is known as Diffraction grating.

Gratings are constructed by ruling equidistant parallel lines on a transparent material such as glass, with a fine diamond point. The ruled lines are opaque to light while the space between any two lines is transparent to light and acts as a slit. This is known as plane transmission grating. When the spacing between the lines is of the order of the wavelength of light, then an appreciable deviation of the light is produced.

A section of a plane transmission grating AB placed perpendicular to the plane of the paper is as shown in the figure.

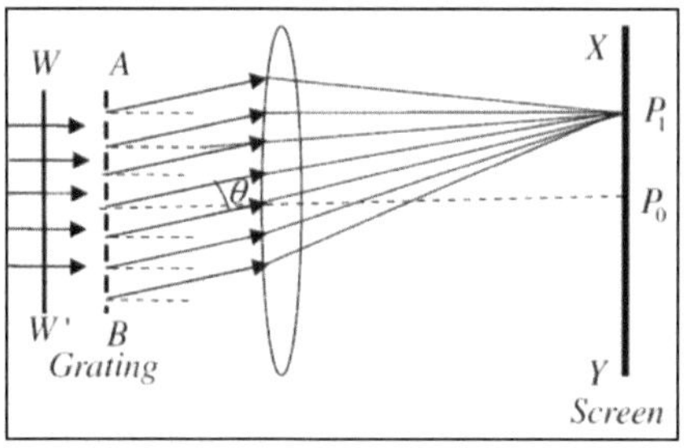

Let 'e' be the width of each slit and 'd' the width of each opaque space. Then (e + d) is known as grating element and XY is the screen. Suppose a parallel beam of monochromatic light of wavelength 'λ' be incident normally on the grating. By Huygen's principle, each of the slit sends secondary wavelets in all directions. Now, the secondary wavelets travelling in the direction of incident light will focus at a point P_0 on the screen. This point P_0 will be a central maximum.

Now consider the secondary waves travelling in a direction inclined at an angle 'θ' with the incident light will reach point P_1 in different phases. As a result dark and bright bands on both sides of central maximum are obtained.

The intensity at point P_1 may be considered by applying the theory of Fraunhofer diffraction at a single slit. The wavelets proceeding from all points in a slit along their direction are equivalent to a single wave of amplitude $A\frac{\sin\alpha}{\alpha}$ starting from the middle point of the slit, Where $\alpha = \frac{\pi}{\lambda} e \sin\theta$.

If there are N slits, then we have N diffracted waves. The path difference between two consecutive slits is $(e+d)\sin\theta$. Therefore, the phase difference,

$$\delta = \frac{2\pi}{\lambda}(e+d)\sin\theta = 2\beta$$

Hence the intensity in a direction 'v' can be found by finding the resultant amplitude of N vibrations each of amplitude $A\frac{\sin\alpha}{\alpha}$ and a phase difference of 2β.

Since in the previous case,

$$a = A\frac{\sin\alpha}{\alpha}; n = N; \delta = 2\beta.$$

Substituting these in equation $R = a\frac{\sin(n\delta/2)}{\sin(\delta/2)}$

The resultant amplitude on screen at P_1 becomes,

$$R = \left(A\frac{\sin\alpha}{\alpha}\right)\frac{\sin N\beta}{\sin\beta}$$

Thus Intensity at P_1 will be,

$$I^2 = R^2 = \left(A\frac{\sin\alpha}{\alpha}\right)^2\frac{\sin^2 N\beta}{\sin^2\beta}$$

The factor,

$$\left(A\frac{\sin\alpha}{\alpha}\right)^2$$

gives the distribution of Intensity due to a single slit while the factor,

$$\frac{\sin^2 N\beta}{\sin^2\beta}$$

gives the distribution of intensity as a combined effect of all the slits.

Intensity Distribution

Case (i): Principal maxima: The equation $R = \left(A\frac{\sin\alpha}{\alpha}\right)\frac{\sin N\beta}{\sin\beta}$ will take a maximum value if,

$$\sin\beta = 0$$

$$\beta = \pm n\pi; n = 0,1,2,3....$$

$$\frac{\pi}{\lambda}(e+d)\sin\theta = \pm n\pi$$

$$(e+d)\sin\theta = \pm n\lambda$$

n = 0 corresponds to zero order maximum. For n = 1, 2, 3,... we obtain first, second, third,... principal maxima respectively. The ± sign indicates that there are two principal maxima of the same order lying on either side of zero order maximum.

Case(ii): Minima Positions: The equation $R=\left(A\frac{\sin\alpha}{\alpha}\right)\frac{\sin N\beta}{\sin\beta}$ takes minimum value if $\sin N\beta=0$ but $\sin\beta\neq 0$,

$$\therefore N\beta=\pm m\pi$$

$$N\frac{\pi}{\lambda}(e+d)\sin\theta=\pm m\pi$$

$$N(e+d)\sin\theta=\pm m\lambda$$

Where m has all integral values except m = 0, N, 2N, ..., nN, because for these values $\sin\beta$ becomes zero and we get principal maxima. Thus, m = 1, 2, 3, ..., (N-1). Hence,

$$N(e+d)\sin\theta=\pm m\lambda \text{ where } m=1,2,3....(N-1),(N+1),....(2N-1)...$$

gives the minima positions which are adjacent to the principal maxima.

Case(iii): Secondary maxima: As there are (N-1) minima between two adjacent principal maxima there must be (N-2) other maxima between two principal maxima. These are known as secondary maxima. To find their positions,

$$\frac{dI}{d\beta}=0$$

$$\frac{dI}{d\beta}=\left(A\frac{\sin\alpha}{\alpha}\right)^2 2\left(\frac{\sin N\beta}{\sin\beta}\right)\left[\frac{N\cos N\beta\sin\beta-Sin\,N\beta\cos\beta}{\sin^2\beta}\right]=0$$

$$\because\frac{\sin\alpha}{\alpha}\neq 0;\ \sin N\beta\neq 0$$

Only,

$$\left[N\cos N\beta\sin\beta-Sin\,N\beta\cos\beta\right]=0$$

$$N\tan\beta=\tan N\beta$$

The roots of the above equation other than those for which $\beta=\pm n\pi$ give the positions of secondary maxima The eqn $N\tan\beta=\tan N\beta$ can be written as,

$$\tan N\beta=\frac{N}{\cot\beta}$$

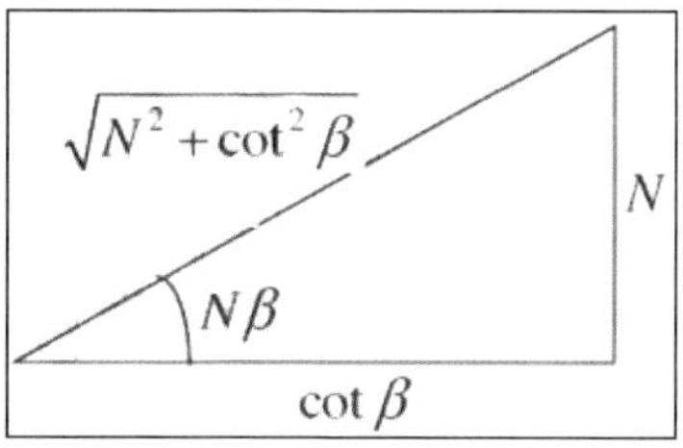

From the triangle we have,

$$\sin N\beta = \frac{N}{\sqrt{N^2 + \cot^2 \beta}}$$

$$\therefore \frac{\sin^2 N\beta}{\sin^2 \beta} = \frac{N^2}{\left(N^2 + \cot^2 \beta\right)\sin^2 \beta}$$

$$\therefore \frac{\sin^2 N\beta}{\sin^2 \beta} = \frac{N^2}{N^2 \sin^2 \beta + \left(1 - \sin^2 \beta\right)}$$

$$\therefore \frac{\sin^2 N\beta}{\sin^2 \beta} = \frac{N^2}{1 + \left(N^2 - 2\right)\sin^2 \beta}$$

$$I = I_o \left(\frac{\sin \alpha}{\alpha}\right)^2 \frac{N^2}{1 + \left(N^2 - 1\right)\sin^2 \beta}$$

Since intensity of principal maxima is proportional to N^2,

$$\frac{\text{Intensity of secondry maxima}}{\text{Intensity of Principal maxima}} = \frac{\dfrac{N^2}{1 + \left(N^2 - 1\right)\sin^2 \beta}}{N^2} = \frac{1}{1 + \left(N^2 - 1\right)\sin^2 \beta}$$

Hence if the value of N is larger, then the secondary maxima will be weaker and becomes negligible when N becomes infinity.

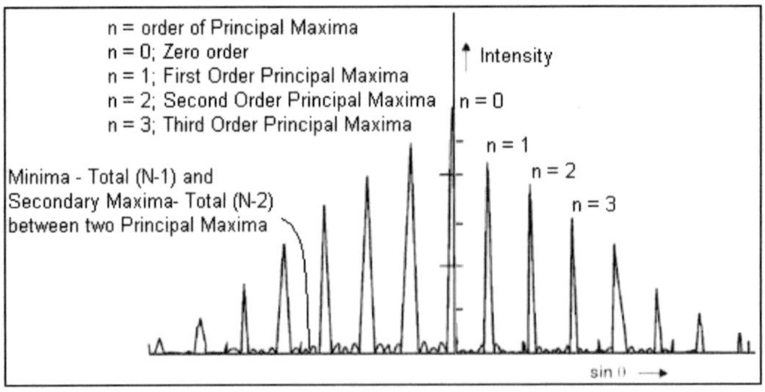

Diffraction Grating

A diffraction grating is an optical device exploiting the phenomenon of diffraction, i.e., an kind of diffractive optics. It contains a periodic structure, which causes spatially varying optical amplitude and/or phase changes.

Diffraction gratings are used to disperse light; that is to spatially separate light of different wavelengths. They have replaced prisms in most fields of spectral analysis.

The manufacture of diffraction gratings dates back to the late 18th century, however extensive use was limited by the difficulties in ruling gratings of adequate quality until Henry Rowland built ruling engines in the latter part of the 19th century. Subsequent production of high quality gratings led to significant advances in analytical spectroscopy. Today, most diffraction gratings are replicated from masters. Masters can be ruled with interferometric control, generated holographically with possible enhancement through ion etching, or made with newer techniques derived from semiconductor lithography.

Plane ruled reflection gratings.

Plane holographic reflection gratings.

The Diffraction Grating Equation

A typical diffraction grating consists of a substrate, usually of an "optical material", with a large number of parallel grooves ruled or replicated in its surface and overcoated with a reflecting material such as aluminum. The quality and spacing of the grooves are crucial to the performance of the grating, but the basic grating equation may be derived by supposing a section through the grating surface, normal to the ruling direction as a sawtooth pattern, shown below in figure.

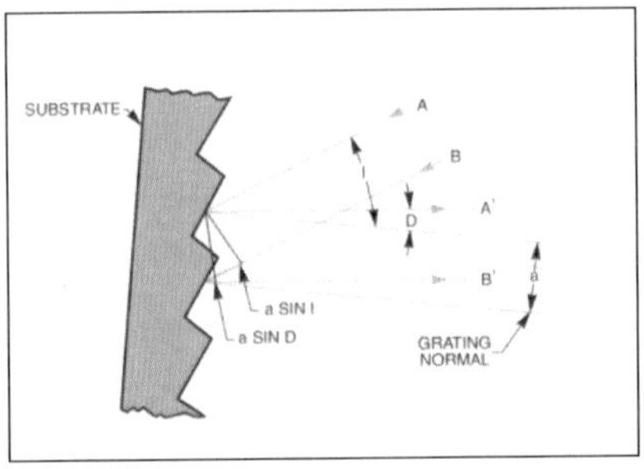

Figure: The sawtooth pattern of a grating section.

Light rays A and B, of wavelength λ, incident on adjacent grooves at angle I to the grating normal are shown. Consider light at angle D to the grating normal; this light originates from the A and B rays as they strike the grating. The path difference between the A^1 and B^1 rays can be seen to be:

$$a \sin I + a \sin D$$

Summing of the rays A^1 and B^1 results in constructive interference if the path difference is equal to any integer multiple of the wavelength λ:

$$a(\sin I + \sin D) = m\lambda$$

Where:

m = an integer, and is the order of diffraction

This is the basic grating equation. Note that if D is on the opposite side of the grating normal from I, it is of the opposite sign. We have considered only two grooves. Including all the other grooves does not change the basic equation but sharpens the peak in the plot of diffracted intensity against angle D.

When a parallel beam of monochromatic light is incident on a grating, the light is diffracted from the grating in directions corresponding to m = -2, -1, 0, 1, 2, 3, etc. This is shown in figure.

When a parallel beam of polychromatic light is incident on a grating then the light is dispersed so that each wavelength satisfies the diffraction grating equation. This is shown in figure.

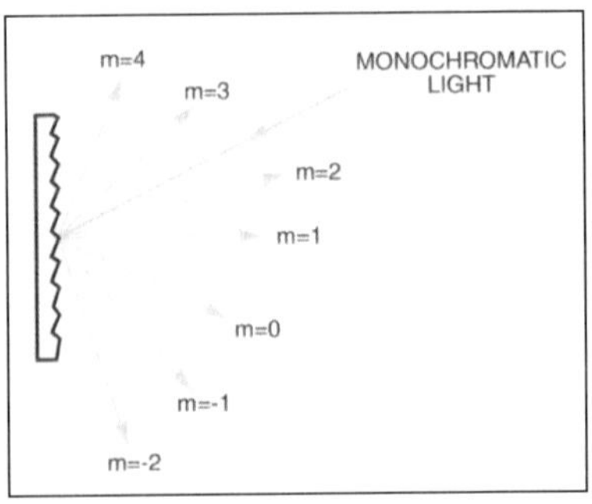

Figure: The "Grating Equation" satisfied for a parallel beam of monochromatic light.

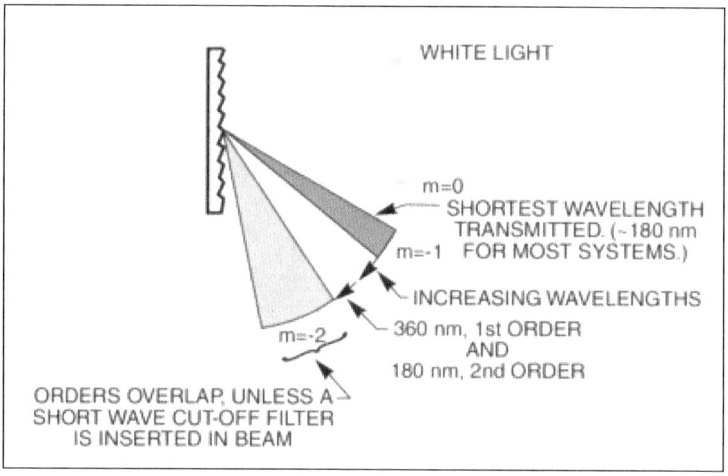

Figure: Polychromatic light diffracted from a grating.
Positive orders have been omitted for clarity.

In most monochromators, the input slit and collimating mirror fix the direction of the input beam that strikes the grating. The focusing mirror and exit slit fix the output direction. Only wavelengths that satisfy the grating equation pass through the exit slit. The remainder of the light is scattered and absorbed inside the monochromator. As the grating is rotated, the angles I and D change, although the difference between them remains constant and is fixed by the geometry of the monochromator. A more convenient form of the grating equation for use with monochromators is:

$$m\lambda = 2\ x.\ x\ \cos\varphi\, x \sin\theta$$

Where:

φ = Half the included angle between the incident ray and the diffracted ray at the grating.

θ = Grating angle relative to the zero order position.

These terms are related to the incident angle I and diffracted angle D by:

$I = \theta + \varphi$ and $D = \theta - \varphi$

Diffraction Grating Order

It is important to note the sign of "m" is given by either form of the grating equation and can be positive or negative. In a monochromator, the angles I and D are determined by the rotational position of the grating. We use the sign convention that all angles that are counter clockwise from the grating normal are positive, and all angles that are clockwise to the grating are negative. The incident light, diffracted light and grating rotation can be at positive or negative angles depending on which side of the grating normal they are. The half angle is always regarded as positive.

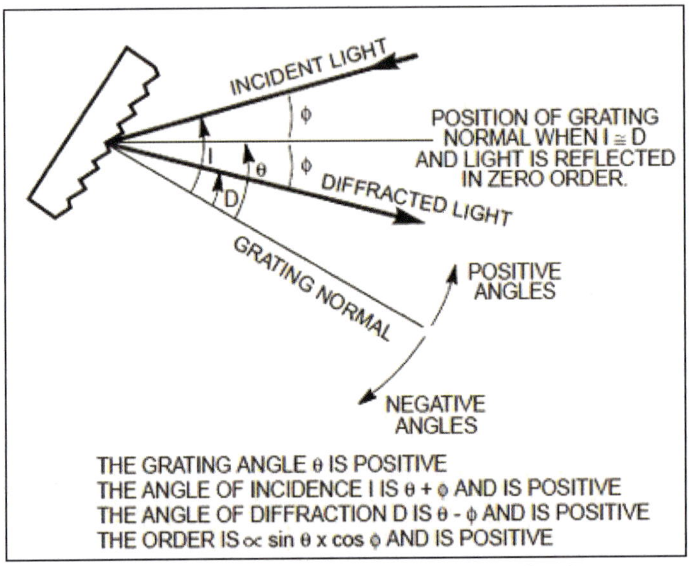

Figure: The sign convention for the angle of incidence, angle of diffraction and grating angle.

If the angle D is equal to I and of opposite sign, then the grating angle and order are zero, and the light is simply being reflected. If the grating angle is positive then the order is positive (m = 1), if the grating angle is negative, then the order is negative (m = −1). Table lists the half angles and the sign of the orders passed by our monochromators.

The grating equation is also satisfied for wavelengths in higher orders, when | m | is >1. Therefore $\lambda_2 = \lambda_1/2$ for m = ±2, $\lambda_3 = \lambda_1/3$ for m = ±3, etc. The wavelength λ_2 is in the second order and λ_3 is in the third order, etc. Again, this concept is illustrated in figure.

Usually only the first order, positive or negative, is desired. The other wavelengths in higher orders may need to be blocked. The input spectrum and detector sensitivity determine whether order sorting or blocking filters are needed.

Blaze Wavelength

If monochromatic light strikes a grating then a fraction of it is diffracted into each order. The fraction diffracted into any order can be termed the efficiency of the grating in that order. All Oriel Gratings are designed for efficient diffraction into the first order. Gratings are not equally "efficient" at all wavelengths for a variety of detailed reasons. The efficiency can be "tuned" by changing the groove facet angles, or shape and depth. The optimization of efficiency by appropriate groove shaping has become known as blazing. The "Blaze Wavelength" is the wavelength for which the blazed grating is most efficient. Figure shows a typical efficiency vs. wavelength curve for a ruled and a blazed holographic grating.

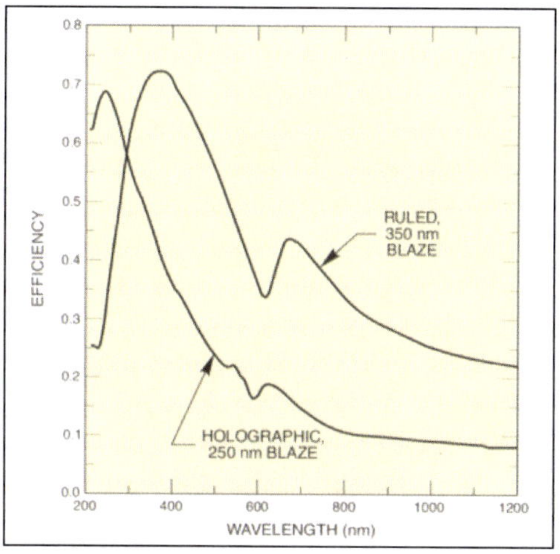

Figure: The efficiency of a ruled and a blazed holographic grating, both have 1200 l/mm.

Holographic Gratings

Holographic gratings are created using a sinusoidal interferometric pattern and sometimes an etching process. Sinusoidally grooved gratings produce very little scattered light but have low, flat efficiency curves, although they are generally quite broad. Blazed holographic gratings use etching during the interferometric process, or an ion gun to form a blaze angle in a secondary process. The former does not produce strong blazing, and while the latter produces high efficiencies at the blaze wavelengths, light scatter is increased due to the formation of micro structure along the edges of the grooves. Newport offers both ruled and holographic gratings to provide the best combination of blaze, efficiency, and low light scatter for different wavelength ranges.

Table: Half angles and orders for oriel monochromators.

Model	Half Angle(degree)	Orders
77250	5.1	Positive
77400	12,5	Positive
77200	13.04	Negative
77700	11.83	Positive
74000	5.1	Positive
74100	11,83	Positive

Dispersion

If we fix the angle I in the grating equation and differentiate with respect to wavelength we get: a x cos D x δD = m x δλ

Thus:

$$\frac{\delta D}{\delta \lambda} = \frac{m}{a \times \cos D}$$

δD/δλ is the angular dispersion or change of diffraction angle corresponding to a small change in wavelength. It is greater for smaller groove spacings, a (greater number of lines per millimeter); larger orders, m; and larger diffraction angles, D. The linear dispersion, δL/δλ, at a monochromator's exit slit will vary with output focal length, f, and angle D and is the product of the focal length and angular dispersion:

$$\frac{\delta L}{\delta \lambda} = f \times \frac{\delta D}{\delta \lambda} = \frac{f \times m}{a \times \cos D}$$

For our 77250 Monochromator and MS257 Monochromator. and Spectrograph each with 1200 l/mm gratings, the linear dispersion is:

$$\overset{77250}{\frac{125 \times m \times 1200}{10^6 \times \cos D}} \text{mm/nm} \quad \overset{77000}{\frac{257 \times m \times 1200}{10^6 \times \cos D}} \text{mm/nm}$$

Normally, we use the "reciprocal linear dispersion", which gives the wavelength dispersion in nm/mm of slit width.

Bandpass and Resolution

The bandpass is the width of the spectrum passed by a monochromator or spectrograph when illuminated by a light source with a continuous spectrum.

In a monochromator, the bandpass may be decreased by reducing the width of the slits until a limiting bandpass is reached. The limiting bandpass is called the resolution of the instrument. In spectral analysis, the resolution is a measure of the ability of the instrument to separate two spectral lines that are close together. The resolution of the 77250 is about 0.5 nm and that of the 77700 is 0.1 nm with 1200 l/mm gratings in the visible. For a properly illuminated grating, the aberrations of the optical system determine the resolution. For any instrument, the optical aberrations are minimized by illuminating only the central zone of the input slit.

For large bandpasses, compared with the resolution, the bandpass is given by the product of the reciprocal linear dispersion and the slit width.

Diffraction Grating Illumination and Resolution

The resolving power R of a grating is defined in terms of wavelengths λ and λ + δλ, where λ + δλ is the closest wavelength to λ that can be resolved. Theoretically:

$$R = \frac{\lambda}{\delta \lambda} = \frac{W(\sin I + \sin D)}{\lambda}$$

Where:

$\delta\lambda$ = Resolution

W = Illuminated width of the grating.

Normally the smallest slit available and optical aberrations, rather than the grating, determine the attainable resolution. For the 77700 at λ = 500 nm, with a 1200 l/mm grating R is 60000 with a fully illuminated grating. Based on this, $\delta\lambda$ is 0.008 nm. With 25 µm slits the product of slitwidth and reciprocal linear dispersion is 0.09 nm, which is close to the measured value of 0.1 nm and R = 5000 instead of 60000. If however, only a few mm of the grating are illuminated, as is sometimes the case with laser sources incorrectly coupled to the monochromator, the grating resolution can broaden the measured bandwidths. For example, if a laser illuminates only 2 mm of the grating width, the measured bandwidth will be 0.2 nm, even for a very narrowband laser.

Fabrication Methods for Gratings

Gratings can be fabricated with different methods:

- A traditional technique is based on a ruling engine, which mechanically imprints the required surface relief (a groove structure) on a metallic surface, for example. Although *ruled gratings* are difficult to make with very small line spacings, they can be used for robust metallic blazed gratings with high diffraction efficiency. A disadvantage for use in grating spectrometers is that they cause substantial amounts of stray light due to surface irregularities.
- *Holographic surface gratings* are made with photolithographic techniques, which allow finer grating structures. Simple holographic gratings have a sinusoidal phase variation and thus a low diffraction efficiency, but they cause only little stray light as their surfaces can be very regular. They can be made in a wide range of hard materials such as silica and various semiconductors, and advanced fabrication techniques can produce carefully controlled structures, such as blazed gratings.
- *Holographic volume gratings* have a periodic refractive index variation within a transparent medium. They can have high diffraction efficiency and low stray light, but can be sensitive to changes of temperature and humidity. Their sensitivity to humidity may be reduced by sealing them with suitable surface layers.

It is also possible to fabricate a diffraction grating on a prism; the combination of a prism and a grating is sometimes called a “grism”. One may choose the parameters such that light at a certain center wavelength gets through the grism without any deflection.

Another possibility is to make a grating on top of a dielectric mirror structure, resulting in a reflecting grating mirror with very high diffraction efficiency.

Applications of Diffraction Gratings

Diffraction gratings are commonly used for spectroscopic dispersion and analysis of light. What makes them particularly useful is the fact that they form a sharper pattern than double slits do. That is, their bright fringes are narrower and brighter while their dark regions are darker. Diffraction gratings are key components of monochromators used, for example, in optical imaging of particular wavelengths from biological or medical samples. A diffraction grating can be chosen to specifically analyze a wavelength emitted by molecules in diseased cells in a biopsy sample or to help excite strategic molecules in the sample with a selected wavelength of light. Another vital use is in optical fiber technologies where fibers are designed to provide optimum performance at specific wavelengths. A range of diffraction gratings are available for selecting wavelengths for such use.

Calculating Typical Diffraction Grating Effects

Diffraction gratings with 10,000 lines per centimeter are readily available. Suppose you have one, and you send a beam of white light through it to a screen 2.00 m away.

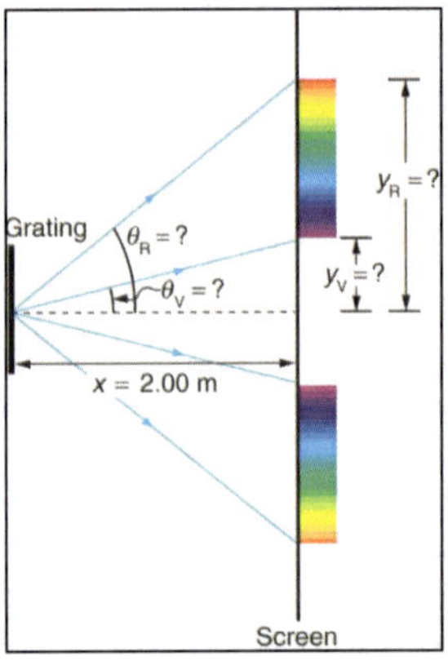

Figure: (a) The diffraction grating considered in this example produces a rainbow of colors on a screen a distance x = 2.00m from the grating. The distances along the screen are measured perpendicular to the x-direction. In other words, the rainbow pattern extends out of the page.

- Find the angles for the first-order diffraction of the shortest and longest wavelengths of visible light (380 and 760 nm, respectively).
- What is the distance between the ends of the rainbow of visible light produced on the screen for first-order interference?
- In a bird's-eye view, the rainbow pattern can be seen on a table where the equipment is placed.

Strategy

Once a value for the diffraction grating's slit spacing d has been determined, the angles for the sharp lines can be found using the equation,

$$d \sin\theta = m\lambda \, form = 0, \pm1, \pm2, \ldots.$$

Since there are 10,000 lines per centimeter, each line is separated by 1/10,000 of a centimeter. Once we know the angles, we an find the distances along the screen by using simple trigonometry.

The distance between slits is $d = (1cm)/10,000 = 1.00 \times 10 \quad cm$ or $1.00 \times 10^{-6} m$. Let us call the two angles θ_V for violet (380 nm) and θ_R for red (760 nm). Solving the equation $d \sin\theta_V = m\lambda$ for $\sin\theta_V$,

$$\sin\theta_V = \frac{m\lambda_V}{d},$$

where m = 1 for the first-order and $\lambda_V = 380nm = 3.80 \times 10^{-7} m$. Substituting these values gives,

$$\sin\theta_V = \frac{3.80 \times 10^{-7} m}{1.00 \times 10^{-6} m} = 0.380.$$

Thus the angle θ_V is,

$$\theta_V = \sin^{-1} 0.380 = 22.33°.$$

Similarly,

$$\sin\theta_R = \frac{7.60 \times 10^{-7} m}{1.00 \times 10^{-6} m} = 0.760.$$

Thus the angle θ_R is, $\theta_R = \sin^{-1} 0.760 = 49.46°$.

Notice that in both equations, we reported the results of these intermediate calculations to four significant figures to use with the calculation in part (b).

The distances on the screen are labeled y_V and y_R. Notice that $\tan\theta = y/x$. We can solve for y_V and y_R. That is,

$$y_V = x\tan\theta_V = (2.00m)(\tan 22.33°) = 0.815\,m$$

and

$$y_R = x\tan\theta_R = (2.00m)(\tan 49.46°) = 2.338m.$$

The distance between them is therefore,

$$y_R - y_V = 1.523m.$$

Significance

The large distance between the red and violet ends of the rainbow produced from the white light indicates the potential this diffraction grating has as a spectroscopic tool.

The more it can spread out the wavelengths (greater dispersion), the more detail can be seen in a spectrum. This depends on the quality of the diffraction grating—it must be very precisely made in addition to having closely spaced lines.

Fresnel Diffraction

In optics, the Fresnel diffraction equation for near-field diffraction is an approximation of the Kirchhoff–Fresnel diffraction that can be applied to the propagation of waves in the near field. It is used to calculate the diffraction pattern created by waves passing through an aperture or around an object, when viewed from relatively close to the object. In contrast the diffraction pattern in the far field region is given by the Fraunhofer diffraction equation.

The near field can be specified by the Fresnel number, F of the optical arrangement. When $F \gg 1$ the diffracted wave is considered to be in the near field. However, the validity of the Fresnel diffraction integral is deduced by the approximations derived below. Specifically, the phase terms of third order and higher must be negligible, a condition that may be written as,

$$\frac{F\theta^2}{4} \ll 1,$$

where θ is the maximal angle described by $\theta \approx a / L$, a and L the same as in the definition of the Fresnel number.

The multiple Fresnel diffraction at closely spaced periodical ridges (ridged mirror) causes the specular reflection; this effect can be used for atomic mirrors.

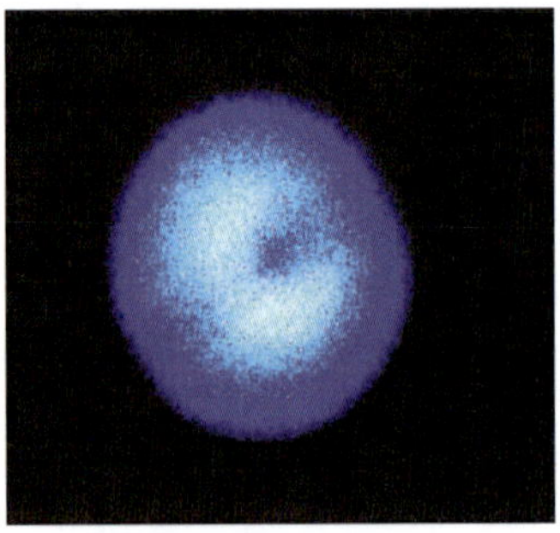

Fresnel diffraction showing central Arago spot.

Early Treatments of this Phenomenon

Some of the earliest work on what would become known as Fresnel diffraction was carried out by Francesco Maria Grimaldi in Italy in the 17th century. In his monograph entitled "Light", Richard C. MacLaurin explains Fresnel diffraction by asking what happens when light propagates, and how that process is affected when a barrier with a slit or hole in it is interposed in the beam produced by a distant source of light. He uses the Principle of Huygens to investigate, in classical terms, what transpires. The wave front that proceeds from the slit and on to a detection screen some distance away very closely approximates a wave front originating across the area of the gap without regard to any minute interactions with the actual physical edge.

The result is that if the gap is very narrow only diffraction patterns with bright centers can occur. If the gap is made progressively wider, then diffraction patterns with dark centers will alternate with diffraction patterns with bright centers. As the gap becomes larger, the differentials between dark and light bands decrease until a diffraction effect can no longer be detected.

MacLaurin does not mention the possibility that the center of the series of diffraction rings produced when light is shone through a small hole may be black, but he does point to the inverse situation wherein the shadow produced by a small circular object can paradoxically have a bright center.

In his *Optics*, Francis Weston Sears offers a mathematical approximation suggested by Fresnel that predicts the main features of diffraction patterns and uses only simple mathematics. By considering the perpendicular distance from the hole in a barrier screen to a nearby detection screen along with the wavelength of the incident light, it is possible to compute a number of regions called half-period elements or Fresnel zones. The inner zone is a circle and each succeeding zone will be a concentric annular ring. If the diameter of the circular hole in the screen is sufficient to expose the first or central Fresnel zone, the amplitude of light at the center of the detection screen will be double what it would be if the detection screen were not obstructed. If the diameter of the circular hole in the screen is sufficient to expose two Fresnel zones, then the amplitude at the center is almost zero. That means that a Fresnel diffraction pattern can have a dark center. These patterns can be seen and measured, and correspond well to the values calculated for them.

The Fresnel Diffraction Integral

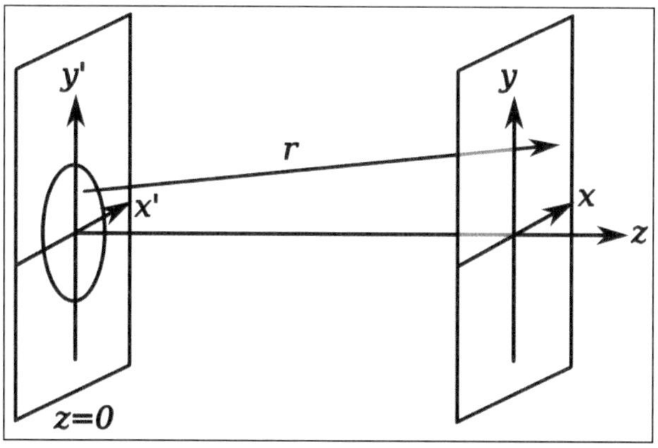

Diffraction geometry, showing aperture (or diffracting object) plane and image plane, with coordinate system.

The electric field diffraction pattern at a point *(x, y, z)* is given by:

$$E(x,y,z)=\frac{1}{i\lambda}\iint_{-\infty}^{+\infty}E(x',y',0)\frac{e^{ikr}}{r}cos(\theta)dx'dy'$$

Where

$E(x',y',0)$ is the electric field at the aperture,

$r=\sqrt{(x-x')^2+(y-y')^2+z^2}$

k is the wavenumber $2\pi/\lambda$

i is the imaginary unit.

Analytical solution of this integral is impossible for all but the simplest diffraction geometries. Therefore, it is usually calculated numerically.

The Fresnel Approximation

The main problem for solving the integral is the expression of r. First, we can simplify the algebra by introducing the substitution:

$$\rho^2=(x-x')^2+(y-y')^2$$

Substituting into the expression for r, we find:

$$r=\sqrt{\rho^2+z^2}=z\sqrt{1+\frac{\rho^2}{z^2}}$$

Next, by the binomial expansion,

$$\sqrt{1+u}=(1+u)^{\frac{1}{2}}=1+\frac{u}{2}-\frac{u^2}{8}+\cdots$$

We can express r as,

$$r=z\sqrt{1+\frac{\rho^2}{z^2}}$$

$$=z\left[1+\frac{\rho^2}{2z^2}-\frac{1}{8}\left(\frac{\rho^2}{z^2}\right)^2+\cdots\right]$$

$$=z+\frac{\rho^2}{2z}-\frac{\rho^4}{8z^3}+\cdots$$

If we consider all the terms of binomial series, then there is no approximation. Let us substitute this expression in the argument of the exponential within the integral; the key to the Fresnel approximation is to assume that the third term is very small and can be ignored and henceforth any higher orders. In order to make this possible, it has to contribute to the variation of the exponential for an almost null term. In other words, it has to be much smaller than the period of the complex exponential; i.e., 2π :

$$k\frac{\rho^4}{8z^3} \ll 2\pi$$

expressing k in terms of the wavelength,

$$k = \frac{2\pi}{\lambda}$$

we get the following relationship:

$$\frac{\rho^4}{z^3\lambda} \ll 8$$

Multiplying both sides by z^3 / λ^3, we have

$$\frac{\rho^4}{\lambda^4} \ll 8\frac{z^3}{\lambda^3}$$

or, substituting the earlier expression for ρ^2,

$$\frac{1}{\lambda^4}\left[(x-x')^2+(y-y')^2\right]^2 \ll 8\frac{z^3}{\lambda^3}$$

If this condition holds true for all values of x, x', y and y', then we can ignore the third term in the Taylor expression. Furthermore, if the third term is negligible, then all terms of higher order will be even smaller, so we can ignore them as well.

For applications involving optical wavelengths, the wavelength λ is typically many orders of magnitude smaller than the relevant physical dimensions. In particular:

$$\lambda \ll z$$

and

$$\lambda \ll \rho$$

Thus, as a practical matter, the required inequality will always hold true as long as,

$$\rho \ll z$$

We can then approximate the expression with only the first two terms:

$$r \approx z + \frac{\rho^2}{2z} = z + \frac{(x-x')^2 + (y-y')^2}{2z}$$

This equation, then, is the Fresnel approximation, and the inequality stated above is a condition for the approximation's validity.

Fresnel Diffraction

The condition for validity is fairly weak, and it allows all length parameters to take comparable values, provided the aperture is small compared to the path length. For the r in the denominator we go one step further, and approximate it with only the first term, $r \approx z$. This is valid in particular if we are interested in the behaviour of the field only in a small area close to the origin, where the values of x and y are much smaller than z. In general, Fresnel diffraction is valid if the Fresnel number is approximately 1.

For Fresnel diffraction the electric field at point (x, y, z) is then given by :

$$E(x,y,z) = \frac{e^{ikz}}{i\lambda z} \iint_{-\infty}^{+\infty} E(x',y',0) e^{\frac{ik}{2z}\left[(x-x')^2+(y-y')^2\right]} dx'dy'$$

This is the Fresnel diffraction integral; it means that, if the Fresnel approximation is valid, the propagating field is a spherical wave, originating at the aperture and moving along z. The integral modulates the amplitude and phase of the spherical wave. Analytical solution of this expression is still only possible in rare cases. For a further simplified case, valid only for much larger distances from the diffraction source, Unlike Fraunhofer diffraction, Fresnel diffraction accounts for the curvature of the wavefront, in order to correctly calculate the relative phase of interfering waves.

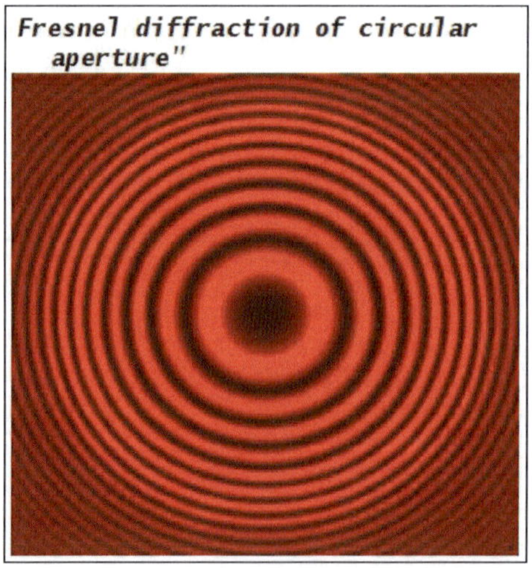

Fresnel diffraction of circular aperture, plotted with Lommel functions.

Alternative Forms

Convolution

The integral can be expressed in other ways in order to calculate it using some mathematical properties. If we define the following function:

$$h(x,y,z)=\frac{e^{ikz}}{i\lambda z}e^{i\frac{k}{2z}\left(x^2+y^2\right)}$$

then the integral can be expressed in terms of a convolution:

$$E(x,y,z)=E(x,y,0)*h(x,y,z)$$

in other words we are representing the propagation using a linear-filter modeling. That is why we might call the function h(x, y, z) the impulse response of free space propagation.

Fourier Transform

Another possible way is through the Fourier transform. If in the integral we express *k* in terms of the wavelength:

$$k=\frac{2\pi}{\lambda}$$

and expand each component of the transverse displacement:

$$\left(x-x'\right)^2=x^2+x'^2-2xx',$$
$$\left(y-y'\right)^2=y^2+y'^2-2yy',$$

then we can express the integral in terms of the two-dimensional Fourier transform. Let us use the following definition:

$$G(p,q)=\mathcal{F}\{g(x,y)\}\equiv\iint_{-\infty}^{\infty}g(x,y)e^{-i2\pi(px+qy)}\,dx\,dy,$$

where p and q are spatial frequencies (wave numbers). The Fresnel integral can be expressed as,

$$E(x,y,z)=\frac{e^{ikz}}{i\lambda z}e^{i\frac{\pi}{\lambda z}\left(x^2+y^2\right)}\mathcal{F}\left\{E(x',y',0)e^{i\frac{\pi}{\lambda z}\left(x'^2+y'^2\right)}\right\}\Bigg|_{p=\frac{x}{\lambda z},\,q=\frac{y}{\lambda z}}$$

$$=H(p,q)\cdot G(p,q)\Big|_{p=\frac{x}{\lambda z},\,q=\frac{y}{\lambda z}},$$

Where,

$$H(p,q) = \mathcal{F}\{h(x,y)\}.$$

That is, first multiply the field to be propagated by a complex exponential, calculate its two-dimensional Fourier transform, replace (p, q) with $\left(\frac{x}{\lambda z}, \frac{y}{\lambda z}\right)$ and multiply it by another factor. This expression is better than the others when the process leads to a known Fourier transform, and the connection with the Fourier transform is tightened in the linear canonical transformation, discussed below.

Linear Canonical Transformation

From the point of view of the linear canonical transformation, Fresnel diffraction can be seen as a shear in the time-frequency domain, corresponding to how the Fourier transform is a rotation in the time-frequency domain.

Fresnel Zone Plate

A suitable circular absorbing or phase grating is called Fresnel zone plate (FZP). A FZP focuses the incoming beam to a point focus. The minimum diameter of the focus reached is 15 nm for X-rays in 2008. There are also zone plates reflecting the incoming light.

These types of Fresnel zone plates can be distinguished:

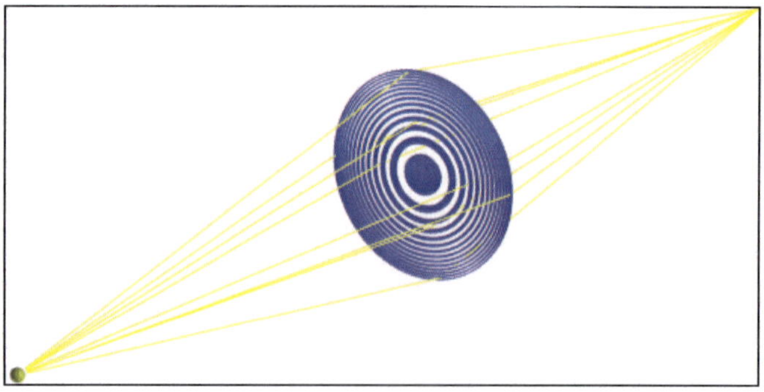

Focusing X-rays with a diffracting zone plate

As for visible light, Fresnel zone plates for X-rays can be absorbing or phase shifting. Absorbing zone plates are more common in the X-ray range. They are made of an X-ray absorbing material, e. g. some micrometers gold, on a membrane transparent for X-rays, e. g. Si_3N_4. Normally the absorber metal is structured by electro deposition into a pre-structured resist. The resist is structured by electron or ion beam writing.

Table: Gives an overview over the principal parameters achieved with Fresnel zone plates in 2009.

Parameter	symbol	Typical values	
		micro zone plate	condenser zone plate
Diameter	D	<0.1 mm	<10 mm
Focal length	f	0.5-1.5 mm	<20 mm
Number of zones	N	100-1000	<50.000
Minimum zone width	dr_n	15-50 nm	30-50 nm
Spatial resolution (Rayleigh criterion)	δ	1.22 dr_n nm	
Diffraction efficiency	η_m	(5...30)%	

As it is technically demanding to produce the narrow outer rings of a zone plate, sometimes so called "composite zone plates" are used. These consist of an inner zone plate used in 1st diffraction order surrounded by an outer zone plate ring with less narrow structures used in 2nd diffraction order. In this way the outermost ring structures of the outer zone plate are less fine and less difficult to produce.

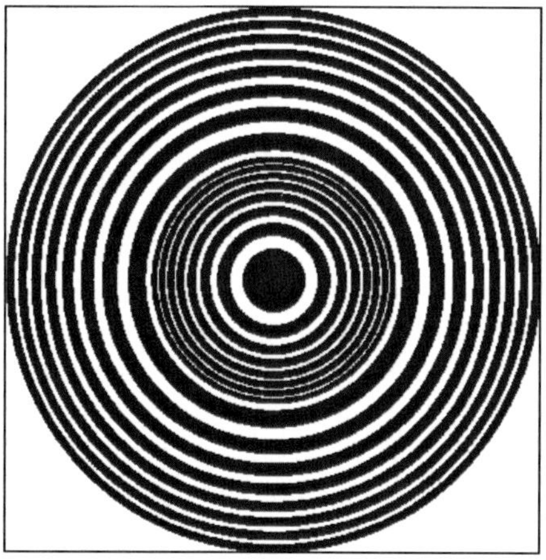

Composite zone plate with inner zone plate and outer zone plate, used in first respectively second diffraction order.

Calculation of Fresnel Zone Plates

The width dr_n of the rings of the circular grating decreases with increasing radius r_n of the rings. The ring radii can be calculated for monochromatic light of the wavelength λ emitted by a point source Q focused to a point P at a distance of $g+b=n\lambda$ with n a natural number, g the object distance and b the image distance. For each optical path from the source through the zone plate to the focus point, the light should interfere constructively. Figure shows the optical path for light hitting the zone plate at a distance r_n from the optical axis.

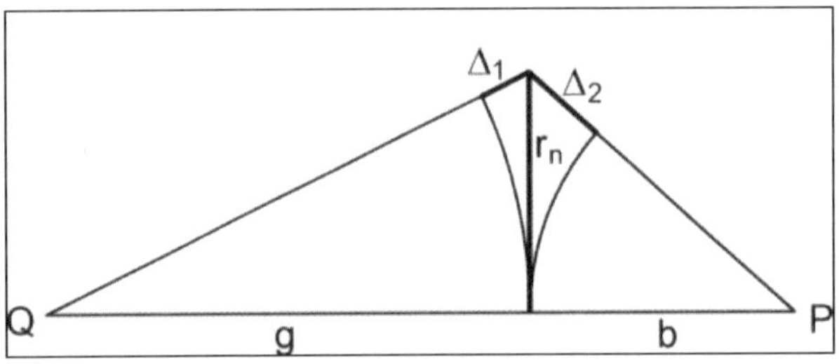

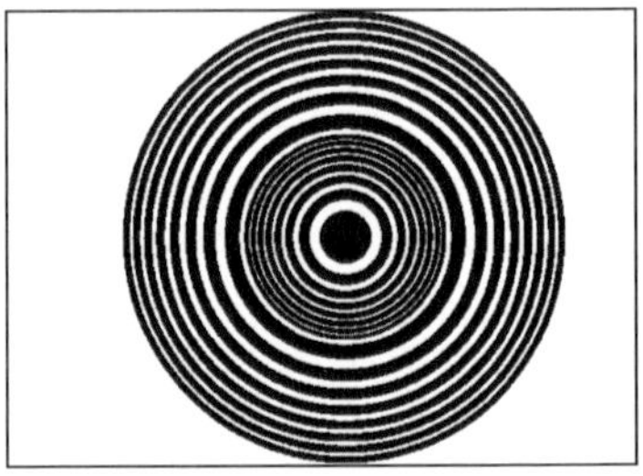

Sketch of the optical path difference between the source point Q and the focus point P at a distance r_n from the optical axis.

The optical path differences are,

$$\Delta_1 = \sqrt{g^2 + r_n^2} - g$$

and

$$\Delta_2 = \sqrt{b^2 + r_n^2} - b$$

When the total optical path difference is,

$$\Delta_{total} = \Delta_1 + \Delta_2 = n\lambda/2 = \sqrt{g^2 + r_n^2} - g + \sqrt{b^2 + r_n^2} - b$$

The light passing neighbour zones will interfere destructively. When now every second zone is absorbing the X-rays, the light passing the transmissive zones of a Fresnel zone plate will have an optical path difference of $n\lambda$ and hence will interfere constructively in the focus point P. Rearranging results in,

$$\sqrt{g^2 + r_n^2} + \sqrt{b^2 + r_n^2} = g + b + n\lambda/2$$

or squared,

$$g^2 + r_n^2 + b^2 + r_n^2 + 2\sqrt{\left(g^2 + r_n^2\right)\left(b^2 + r_n^2\right)} = \left(n\lambda/2 + g + b\right)^2$$

After rearranging, squaring and separating r_n the radius of the nth Fresnel zone is,

$$r_n^2 = \frac{n^4\lambda^4/16 + n^3\lambda^3(g+b)/2 + n^2\lambda^2\left(gb + (g+b)^2\right) + 4n\lambda gb(g+b)}{4(g+b)\left(n\lambda + (g+b)\right) + n^2\lambda^2}$$

As λ is small compared to b and g, neglecting all higher order terms in λ and using nλ << g + b is a good approximation,

$$r_n^2 \approx n\lambda \frac{gb}{g+b} = n\lambda f$$

So the focal length of a Fresnel zone plate depends on the wavelength. In the last step the equation for the focal length f of a thin lens was used,

$$\frac{1}{f} = \frac{1}{g} + \frac{1}{b} \Leftrightarrow f = \frac{gb}{g+b}$$

Differentiating the approximated r_n with respect to n gives the width dr_n of the nth ring,

$$dr_n = \frac{r_n}{2n}$$

The surface of each zone is constant over the zone plate, when the wavelength and the focal length are fixed. This means that each zone contributes to the focus intensity with an equal amount of transmitted light,

$$F_{n^{th} zone} = \pi\left(r_{n+1}^2 - r_n^2\right) = \pi\lambda f.$$

A grating mostly has more than one diffraction order. In a Fresnel zone plate this leads to more than one focus point. The negative diffraction orders even result in diverging beams, in other words: zone plates also behave like a dispersive lens with virtual focus points. When a zone plate is used as focusing lens, normally only the first diffraction order is used and all other orders have to be blocked by a suitable aperture.

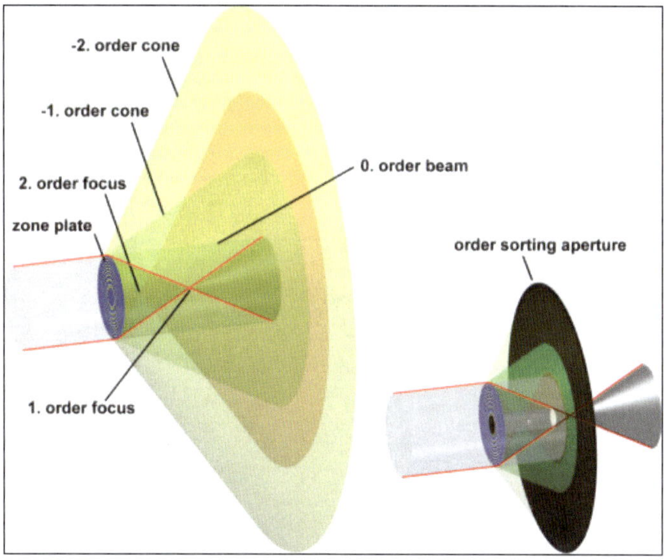

Diffraction orders of a circular grating (left); with order sorting aperture (right); first order rays marked red on mouse over.

Efficiency of Fresnel Zone Plates

Zone plates with absorbing gratings show about two times more absorption than zone plates using phase gratings. Additionally, light passing into unused orders is lost. The diffraction efficiency η_m of a zone plate used in the mth diffraction order is defined as,

$$\eta_m = \frac{I_m}{I_0}$$

with the incoming intensity I_0 and the intensity in the mth diffraction order I_m. Efficiency values of 5 to 30% for $m = 1$ are technically reached.

A zone plate with absorbing gratings reaches the theoretically possible diffraction efficiency only when the thickness of the absorbing structures is sufficient to absorb nearly 100% of the incoming light. When gold is used as absorber, and e. g. 90% absorption should be reached, the necessary thickness of the structures depends on the photon energy.

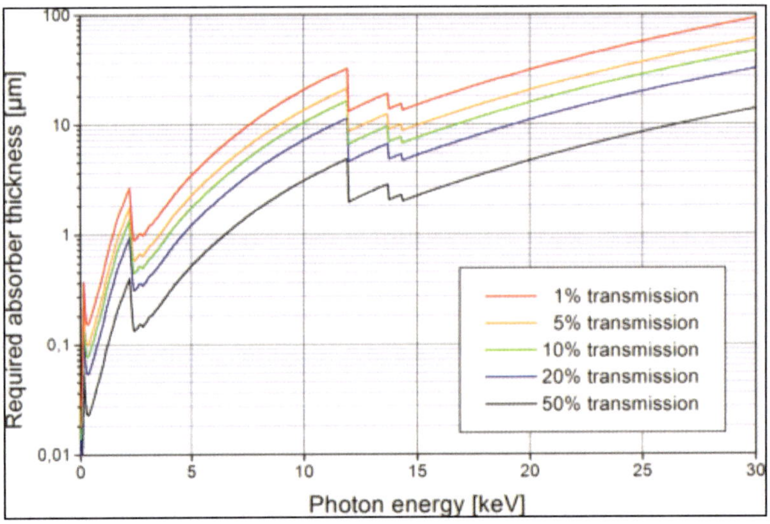

Gold thickness required to absorb most of the incoming light in the zone plate rings.

Spatial Resolution of Fresnel Zone Plates and Fabrication Limitations

The spatial resolution δ_{ZP} of a Fresnel zone plate depends, according to the Abbe theory, on the numerical entrance aperture $NA_{E,\,ZP}$ of the zone plate,

$$NA_{E,ZP} = n_{Material} \sin\left(\Theta_E\right) \approx 1 \cdot \tan\left(\Theta_E\right) = \frac{r_n}{f} = \frac{m\lambda}{2dr_n}$$

with the refractive index decrement n_{Material} (= real part of the refractive index n^*) of the medium between the object and the zone plate, the entrance acceptance angle Θ_E of the zone plate and the diffraction order m used. The following approximations were made: normally the entrance acceptance angle Θ_E is small and the refractive index decrement nearly one.

Based on the Rayleigh-criterion (resolution of two neighbour points resolved with a lens with the numerical aperture *NA*) the largest theoretical possible spatial resolution δ_{ZP} can be calculated for incoherent, monochromatic illumination to be,

$$\delta_{ZP} = 1.22 \frac{\lambda}{2NA_{E,ZP}} = 1.22 \frac{dr_n}{m}$$

In the fabrication process of zone plates the gold absorbers are mostly produced by electroplating a micro structured substrate. The aspect ratio of these micro structures is defined as the ratio of structures height to the smallest lateral structure size. The highest aspect ratio realized in 2009 was in the region of about 120 using the LIGA-process. If the permitted transmission of the absorbing gold structures is relatively high, e.g. 50%, and the aspect ratio achieved in the process is e. g. 100, zone plate for 10 keV light with a minimum spot size of about 40 nm are practicable. To achieve smaller spot sizes or the same spot size for larger photon energies either the required aspect ratio would have to be increased or the efficiency of the zone plate will decrease due to the increasing transmission of the absorber structures.

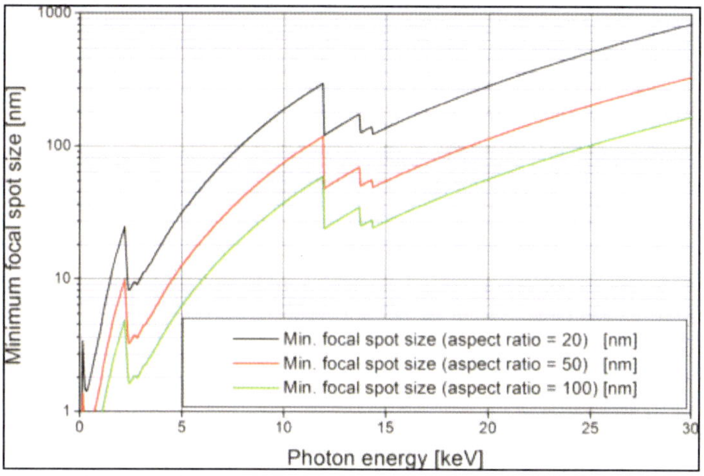

Theoretically minimum achievable spot size when 50% transmission of the gold absorbers is tolerated.

Reflexion Fresnel Zone Plates

Table gives an overview over the principal parameters achieved with reflexion Fresnel zone plates in 2009.

Parameter	symbol	unit	Typical values	
			TER reflexion zone plate	Bragg-Fresnel-reflexion zone plate
Object Distance	g	mm	$\infty - 0.1$	$\infty - 3$

Image Distance	b	mm	$0.1-\infty$	$3-\infty$
Photon Energy Range	E_{Photon}	keV	0.01-1	1-100
Angle of Incidence	Θ_{ZP}	rad	0.001-0.25	0.1-pi/2
Minimal Zone Width	Δ_{ZP}	nm	1000	200
Horizontal Numerical Aperture	$NA_{horizontal}$	rad	0.01-0.25	0.01-0.25
Vertical Numerical Aperture	$NA_{vertical}$	rad	0.1-0.9	0.01-0.25
Focal Spot Diameter	d_U	μm	1-10	0.1-10
Intensity Increase	K	1	5-100	5-1000

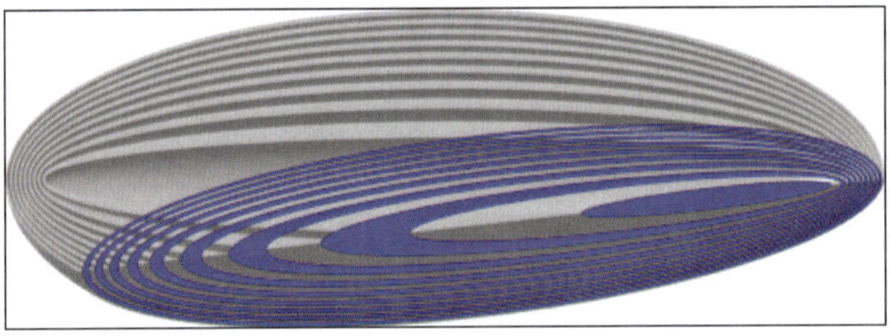

Principle of a reflexion Fresnel zone plate.

Condenser Zone Plates

Special zone plates are designed in order to be used as condensers illuminating the sample in full-field transmission x-ray microscopes. In this case a homogeneously and brightly illuminated square on the sample is required. The distance b from the condenser to the illuminated field, the size of the condenser and the size of the illuminated field D_{zone} have to be matched to the numerical aperture and the field of view of the microscope optics. A suitable condenser design is shown in figure.

Sketch of a condenser zone plate illuminating a square field.

A square zone plate (e. g. 1 mm²) consists of many square gratings with the size of the required field of view (e. g. 20 x 20 fields, each 50 x 50 µm²). All grating grooves in one field are parallel and are oriented perpendicular to the line from the centre of the respective field to the centre of the zone plate (length of the line is L_{zone}). The grating constant g_{zone} of each individual field, for a zone plate operating in first diffraction order, is chosen to be,

$$g_{zone} = \lambda / \sin\gamma$$

With,

$$\gamma = \alpha + \beta = \arctan\frac{L_{zone}}{g} + \arctan\frac{L_{zone}}{b}$$

Under these conditions, the light passing any field will be diffracted towards the field of view on the sample.

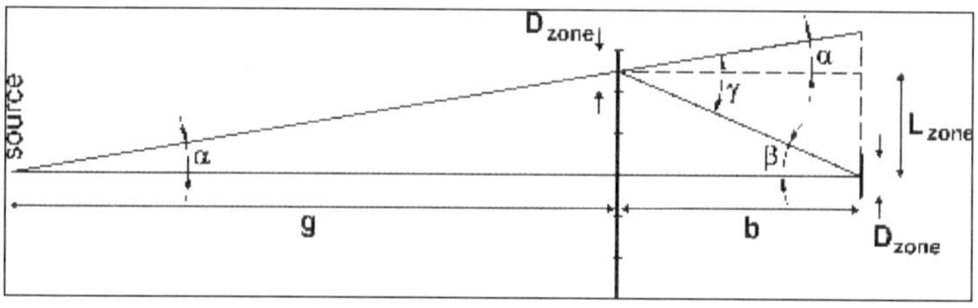

Angles in the calculation of a condenser zone plate.

Gaussian Beam

In optics and particularly in laser physics, laser beams often occur in the form of Gaussian beams, which are named after the mathematician and physicist Johann Carl Friedrich Gauß. Here, the transverse profile of the optical intensity of the beam with a power P can be described with a Gaussian function:

$$I(r,z) = \frac{P}{\pi w(z)^2/2}\exp\left(-2\frac{r^2}{w(z)^2}\right)$$

Where the beam radius $w(z)$ is the distance from the beam axis where the intensity drops to $1/e^2$ (≈ 13.5%) of the maximum value. A hard aperture with radius w can transmit ≈ 86.5% of the optical power. For an aperture radius of $1.5w$ or $2w$, this fraction is increased to 98.9% and 99.97%, respectively.

Note that the factor 1/2 in the denominator in the equation is unfortunately often forgotten, so that the on-axis intensity of the beam is underestimated by a factor of 2.

For example, quoted numbers for the measured damage threshold of optical components are often affected by that problem; the peak intensity at the damage threshold in terms of optical power may have be calculated with or without the mentioned factor, so that a substantial quantitative uncertainty remains for the reader.

The full width at half maximum (FWHM) of the intensity profile is ≈1.18 times the Gaussian beam radius $w(z)$.

In addition to the Gaussian shape of the intensity profile, a Gaussian beam has a transverse phase profile which can be described with a polynomial of at most second order. A linear phase variation in one direction describes a tilt, and a quadratic phase variation is associated with divergence or convergence of the beam.

Gaussian Beam Propagation

In many laser optics applications, the laser beam is assumed to be Gaussian with an irradiance profile that follows an ideal Gaussian distribution. All actual laser beams will have some deviation from ideal Gaussian behavior. The M^2 factor , also known as the beam quality factor, compares the performance of a real laser beam with that of a diffraction-limited Gaussian beam. Gaussian irradiance profiles are symmetric around the center of the beam and decrease as the distance from the center of the beam perpendicular to the direction of propagation increases. This distribution is described by equation:

$$I(r) = I_0 \exp\left(\frac{-2r^2}{w(z)^2}\right) = \frac{2P}{\pi w(z)^2} \exp\left(\frac{-2r^2}{w(z)^2}\right)$$

In the above equation, I_0 is the peak irradiance at the center of the beam, r is the radial distance away from the axis, w(z) is the radius of the laser beam where the irradiance is $1/e^2$ (13.5%) of I_0, z is the distance propagated from the plane where the wavefront is flat, and P is the total power of the beam.

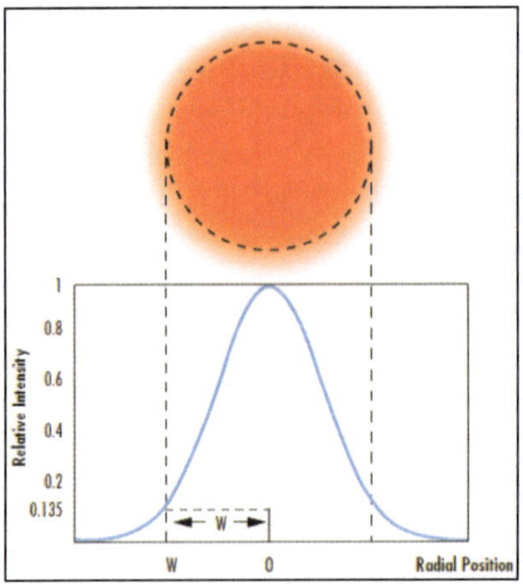

The waist of a Gaussian beam is defined as the location where the irradiance is $1/e^2$ (13.5%) of its maximum value

However, this irradiance profile does not stay constant as the beam propagates through space, hence the dependence of w(z) on z. Due to diffraction, a Gaussian beam will converge and diverge from an area called the beam waist (w_o), which is where the beam diameter reaches a minimum value. The beam converges and diverges equally on both sides of the beam waist by the divergence angle θ. The beam waist and divergence angle are both measured from the axis and their relationship can be seen in the following equations:

$$w_0 = \frac{\lambda}{\pi\theta}$$

$$\theta = \frac{\lambda}{\pi w_0}$$

In the above equations, λ is the wavelength of the laser and θ is a far field approximation. Therefore, θ does not accurately represent the divergence of the beam near the beam waist, but it becomes more accurate as the distance away from the beam waist increases. As seen in equation $\theta = \frac{\lambda}{\pi w_0}$, a small beam waist results in a larger divergence angle, while a large beam waist results in a smaller divergence angle (or a more collimated beam). This explains why beam expanders can reduce beam divergence by increasing beam diameter.

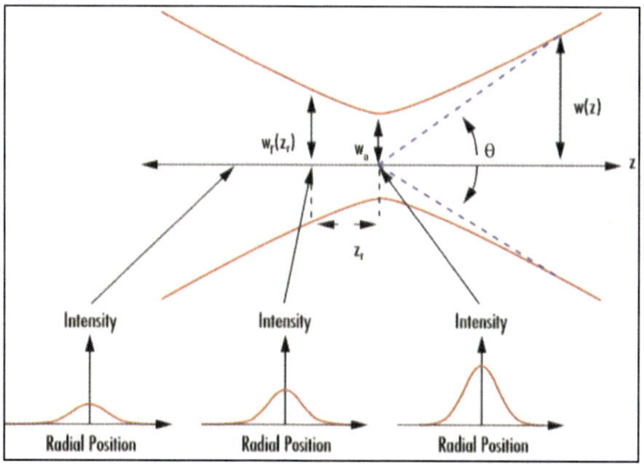

Gaussian beams are defined by their beam waist (w_o), Rayleigh range (z_R), and divergence angle (θ)

Variation of the beam diameter in the beam waist region is defined by:

$$w(z)^2 = w_0^2 + \left(\frac{\lambda}{\pi w_0^2}\right)^2 z^2$$

$$w(z) = w_0\sqrt{1+\left(\frac{\lambda z}{\pi w_0^2}\right)^2}$$

The Rayleigh range of a Gaussian beam is defined as the value of z where the cross-sectional area of the beam is doubled. This occurs when w(z) has increased to $\sqrt{2}w_0$. Using Equation $w(z)=w_0\sqrt{1+\left(\frac{\lambda z}{\pi w_0^2}\right)^2}$, the Rayleigh range ($z_R$) can be expressed as:

$$z_R=\frac{\pi w_0^2}{\lambda}$$

This allows w(z) to also be related to z_R:

$$w(z)=w_0\sqrt{1+\left(\frac{\lambda z}{\pi w_0^2}\right)^2}=w_0\sqrt{1+\left(\frac{z}{z_R}\right)^2}$$

The wavefront of the laser is planar at the beam waist and approaches that shape again as the distance from the beam waist region increases. This occurs because the radius of curvature of the wavefront begins to approach infinity. The radius of curvature of the wavefront decreases from infinity at the beam waist to a minimum value at the Rayleigh range, and then returns to infinity when it is far away from the laser; this is true for both sides of the beam waist.

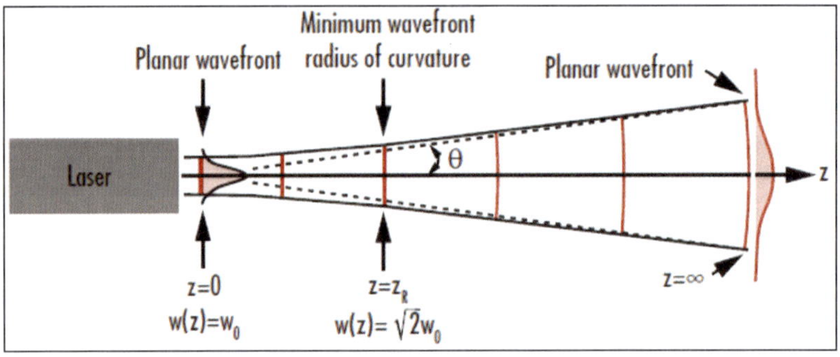

The curvature of the wavefront of a Gaussian beam is near-zero when it is both very close and very far away from the beam waist.

Gaussian Beam Manipulation

Many laser optics systems require manipulation of a laser beam as opposed to simply using the "raw" beam. This may be done using optical components such as lenses, mirrors, prisms, etc. Below is a guide to some of the most common manipulations of Gaussian beams.

The Thin Lens Equation for Gaussian Beams

The behavior of an ideal thin lens can be described using the following equation:

$$\frac{1}{s'}=\frac{1}{s}+\frac{1}{f}$$

In Equation, s' is the distance from the lens to the image, s is the distance from the lens to the object, and f is the focal length of the lens. If the object and image are at opposite sides of the lens, s is a negative value and s' is a positive value. This equation ignores the thickness of a real lens and is therefore only a simple approximation of real behavior. The thin lens equation can also be written in a dimensionless form by multiplying both sides of the equation by f:

$$\frac{f}{s'} = \frac{f}{s} + 1 \text{ or } \frac{1}{\left(\frac{s'}{f}\right)} = \frac{1}{\left(\frac{s}{f}\right)} + 1$$

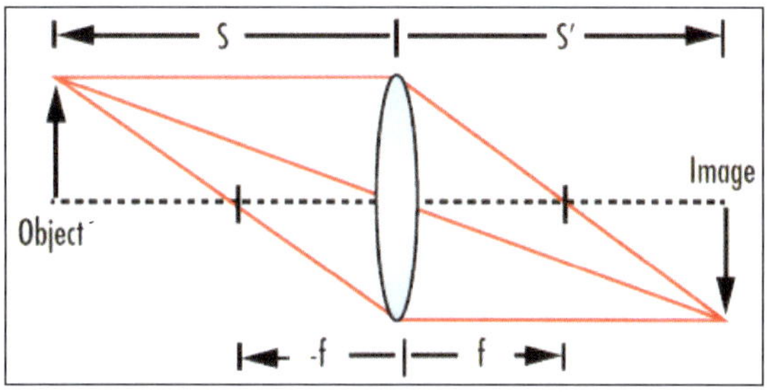

The thin lens equation allows the position of an image (s') to be determined when the distance from the lens to the object (s) and the focal length of the lens (f) are known.

In addition to describing imaging applications, the thin lens equation is applicable to the focusing of a Gaussian beam by treating the waist of the input beam as the object and the waist of the output beam as the image. Gaussian beams remain Gaussian after passing through an ideal lens with no aberrations. In 1983, Sidney Self developed a version of the thin lens equation that took Gaussian propagation into account:

$$\frac{1}{s'} = \frac{1}{s + \frac{z_R^2}{(s+f)}} + \frac{1}{f}$$

The total distance from the laser to the focused spot is calculated by adding the absolute value of s to s'. The following equation can also be written in a dimensionless form by multiplying both sides by f:

$$\frac{f}{s'} = \frac{1}{\left(\frac{s}{f}\right) + \frac{\left(\frac{z_R}{f}\right)^2}{\left(\frac{s}{f} + 1\right)}} + 1$$

This equation approaches the standard thin lens equation as zR/f approaches 0, allowing the standard thin lens equation to be used for lenses with a long focal length. Equations

$$\frac{1}{s'} = \frac{1}{s + \frac{z_R^2}{(s+f)}} + \frac{1}{f}$$ and $$\frac{f}{s'} = \frac{1}{\left(\frac{s}{f}\right) + \frac{\left(\frac{z_R}{f}\right)^2}{\left(\frac{s}{f}+1\right)}} + 1$$ can be used to find the location of the beam waist after being imaged through the lens.

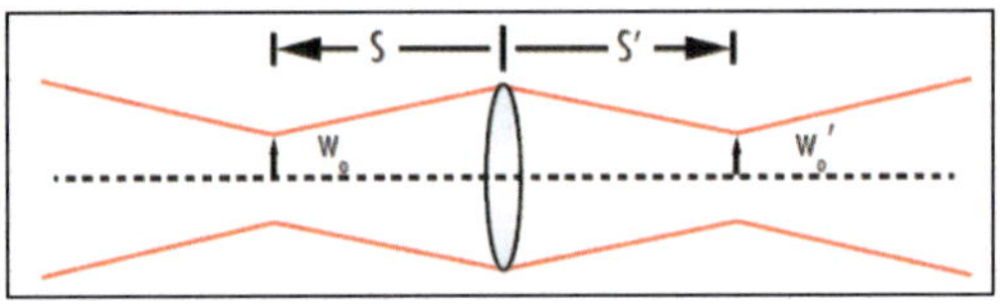

The "object" when refocusing a Gaussian beam is the input waist and the "image" is the output waist.

A plot of the normalized image distance (s'/f) versus the normalized object distance (s/f) shows the possible output waist locations at a given normalized Rayleigh range (z_R/f). This plot shows that Gaussian beams focused through a lens have a few key differences when compared to conventional thin lens imaging. Gaussian beam imaging has both minimum and maximum possible image distances, while conventional thin lens imaging does not. The maximum image distance of a refocused Gaussian beam occurs at an object distance of -(f + z_R), as opposed to –f. The point on the plot where s/f is equal to -1 and s'/f is equal to 1 indicates that the output waist will be at the back focal point of the lens if the input is at the front focal point of a positive lens.

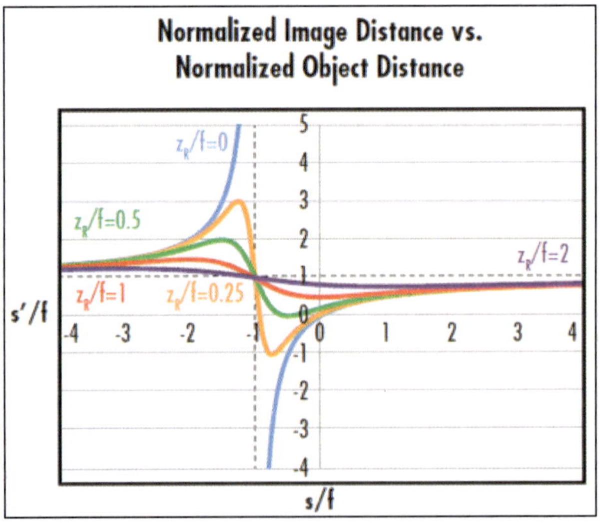

The curve where $z_R/f=0$ corresponds to the conventional thin lens equation. The curves where $z_R/f>0$ show that Gaussian imaging has minimum and maximum image distances defined by the Rayleigh range.

In order to understand the beam waist and Rayleigh range after the beam travels through the lens, it is necessary to know the magnification of the system (α), given by:

$$\alpha = \frac{w_0'}{w_0} = \frac{1}{\sqrt{\left(1-\frac{s}{f}\right)^2 + \left(\frac{z_R}{f}\right)^2}}$$

Where wo is beam waist before the lens and wo' is the beam waist after the lens. The thin lens equation for Gaussian beams can then be rewritten to include the Rayleigh range of the beam after the lens (zR'):

$$\frac{1}{s' + \frac{z'^2_R}{(s'-f)}} = \frac{1}{s} + \frac{1}{f}$$

The above equation will break down if the lens is at the beam waist (s = 0). The inverse of the squared magnification constant can be used to relate the beam waist sizes and locations:

$$\frac{1}{\alpha^2} = \frac{|s|-f}{s'-f} = \left(\frac{w_0}{w_0'}\right)^2$$

Focusing a Gaussian Beam to a Spot

In many applications, such as laser materials processing or surgery, it is highly important to focus a laser beam down to the smallest spot possible to maximize intensity and minimize the heated area. In cases such as these, the goal is to minimize wo'. A modified version of Equation $\frac{1}{\alpha^2} = \frac{|s|-f}{s'-f} = \left(\frac{w_0}{w_0'}\right)^2$ may be used to identify how to minimize the output beam waist:

$$\frac{(|s|-f)^2}{f^2 - \left(\frac{w'_0}{\theta}\right)^2} = \left(\frac{w_0}{w'_0}\right)^2$$

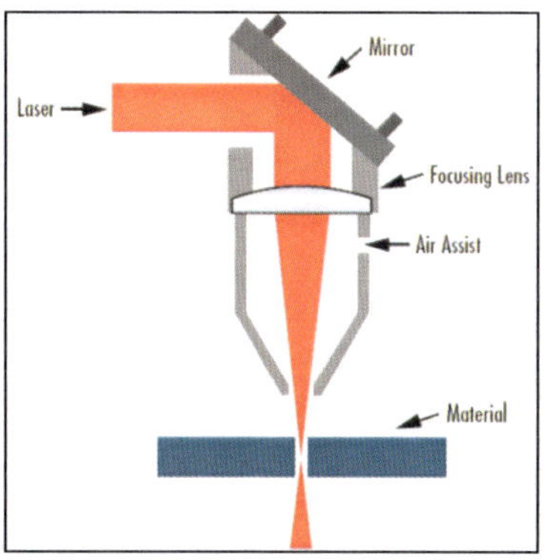

Focusing a laser beam down to the smallest possible size is crucial for a wide range of applications including this laser cutting setup.

After multiplying both sides by the denominator from the left side of the equation and then multiplying both sides by (wo')2, Equation $\frac{\left(|s|-f\right)^2}{f^2-\left(\frac{w'_0}{\theta}\right)^2}=\left(\frac{w_0}{w'_0}\right)^2$ becomes:

$$\left(|s|-f\right)^2=f^2\left(\frac{w_0}{w_0^{'}}\right)^2-\left(\frac{w_0}{\theta}\right)^2=f^2\left(\frac{w_0}{w_0^{'}}\right)^2-z_R^2$$

$$\left(w_0^{'}\right)^2\left[\left(|s|-f\right)^2+z_R^2\right]=f^2w_0^2$$

Solving for w_o' results in:

$$w_0^{'}=w_0\frac{f}{\sqrt{\left(|s|-f\right)^2+z_R^2}}=\alpha w_0$$

$$\alpha=\frac{f}{\sqrt{\left(|s|-f\right)^2+z_R^2}}$$

The focused beam waist can be minimized by reducing the focal length of the lens and |s|-f. The terms next to wo in Equation $w_0^{'}=w_0\frac{f}{\sqrt{\left(|s|-f\right)^2+z_R^2}}=\alpha w_0$ are defined as another form of the magnification constant α in order to compare the values of the input beam to the output beam after going through the lens.

$$\theta'=\frac{\lambda}{\pi w_0^{'}}=\frac{\lambda}{\pi\alpha w_0}=\frac{\theta}{\alpha}$$

$$z_R^{'}=\frac{w_0^{'}}{\theta'}=\frac{\alpha w_0}{\left(\frac{\theta}{\alpha}\right)}=\alpha^2z_R$$

$$s'=f+\alpha^2\left(|s|-f\right)$$

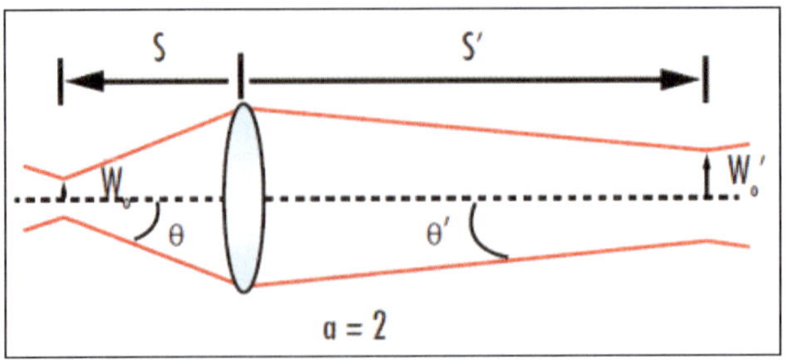

For a magnification of 2, the output beam waist will be twice the input beam waist and the output divergence will be half of the input beam divergence.

There are two limiting cases which further simplify the calculations of the output beam waist size and location: when s is much less than z_R or much greater than ${z_R}^3$. When the lens is well within the laser's Rayleigh range, then $s << z_R$ add $(|s|-f)^2 < [z\_R]^2$. Equation $\alpha = \frac{f}{\sqrt{(|s|-f)^2 + z_R^2}}$ simplifies to:

$$\alpha = \frac{f}{z_R} = \frac{f\theta}{w_0}$$

This also simplifies the calculations for the output beam's waist, divergence, Rayleigh range, and waist location:

$$\theta' = \frac{\theta}{\alpha} = \frac{\theta w_0}{f\theta} = \frac{w_0}{f}$$

$$z_R' = \alpha^2 z_R = \frac{f^2}{z\frac{2}{R}} z_R = \frac{f^2}{z_R}$$

$$w_0' = \alpha w_0 = f\theta$$

$$s' = f + \alpha^2 (|s| - f) = f + \frac{|s| f^2}{z_R^2} \approx f$$

When $s >> z_R$, the distance from the lens to the focused spot equals the focal length of the lens.

The other limiting situation where the lens is far outside of the Rayleigh range and $s >> z_R$, simplifying Equation $\alpha = \frac{f}{\sqrt{(|s|-f)^2 + z_R^2}}$ to:

$$\alpha = \frac{f}{|s|}$$

Which makes the output beam waist diameter:

$$w_0' = \frac{fw}{|s|}$$

Similarly to when $s << z_R$, the calculations for the output beam waist, divergence, Rayleigh range, and beam waist location are also simplified:

$$\theta' = \frac{\theta}{\alpha} = \frac{|s|\theta}{f}$$

$$z_R' = \alpha^2 z_R = \frac{f^2 z_R}{|s|}$$

$$w_0' = \alpha w_0 = \frac{f w_0}{|s|}$$

$$s' = f + \alpha^2\left(|s| - f\right) = f + \frac{\left(|s| - f\right) f^2}{s^2} = f + \frac{f^2}{|s|} \approx f$$

When $s >> z_R$, the distance from the lens to the focused spot equals the focal length of the lens.

Both of these results intuitively make sense because the beam's wavefront is approximately planar both at and very far away from the beam waist. At these locations, the beam is almost perfectly collimated. According to the standard thin lens equation, a collimated input would have an image distance equal to the focal length of the lens.

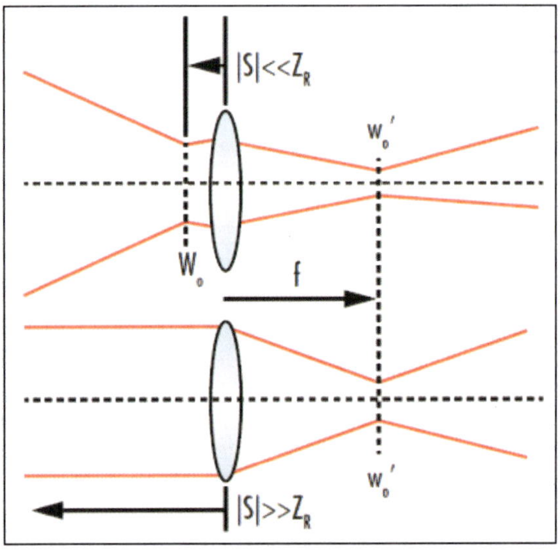

Figure: The focused spot of a Gaussian beam after it passes through a lens will be located at the focal point of the lens if the input beam waist is either very close or very far away from the lens. This is because the input beam is approximately collimated at those points.

Gaussian Focal Shift

Counterintuitively, the intensity of a focused beam in a target at a fixed distance (L) away from the lens is not maximized when the waist is located at the target. The intensity on the target is actually maximized when the waist occurs at a location before the target. This phenomenon is known as Gaussian focal shift.

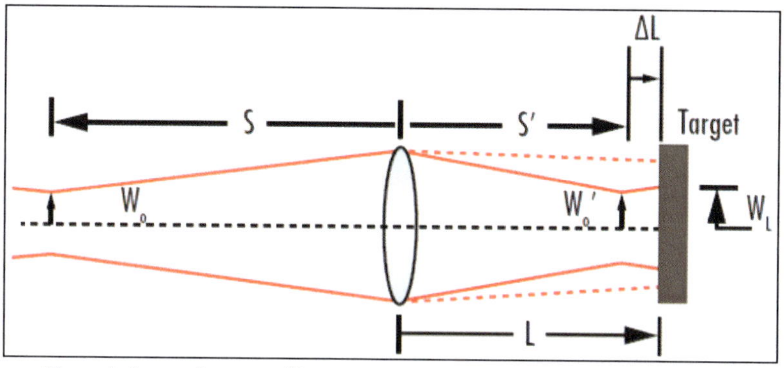

The minimum beam radius at a target occurs when the waist of the focused beam occurs at a specific location before the target, not when focused waist is located at the target.

The beam radius at the target can be described by the following expression:

$$w_L = \frac{w_0}{z_R}\left[f^2 - 2\left(|s| - f\right)\left(L - f\right) + \left(\frac{L-f}{\alpha}\right)^2 \right]$$

Differentiating Equation 34 with respect to the focal length of the focusing lens (f) and solving for f when $d/df\left[w_L(f)\right] = 0$ reveals the lens focal length resulting in the minimum beam radius, and therefore highest intensity, at the target.

$$\frac{d}{df}\left[w_L(f)\right] = \frac{w_0}{z_R}\left[2f + 2\left(|s| - f\right) + \left(L - f\right) - \frac{2\left(L-f\right)}{\alpha^2} \right]$$

$$\frac{d}{df}\left[w_L(f)\right] = 0 \ \text{ when } f = \left(\frac{1}{|s| + \frac{z_R^2}{|s|}} + \frac{1}{L} \right)^{-1}$$

As |s| approaches either zero or infinity, $d/df\left[w_L(f)\right] = 0$ when f=L. In both of these cases, the input beam is approximately collimated, and it thereby follows that the smallest beam radius would occur at the focal point of the lens.

Collimating a Gaussian Beam

Achieving a truly collimated beam where the divergence is 0 is not possible, but achieving an approximately collimated beam by either minimizing the divergence or maximizing the distance between the point of observation and the nearest beam waist is possible. Since the output divergence is inversely proportional to the magnification constant α, the output divergence reaches a minimum value when |s|= f.

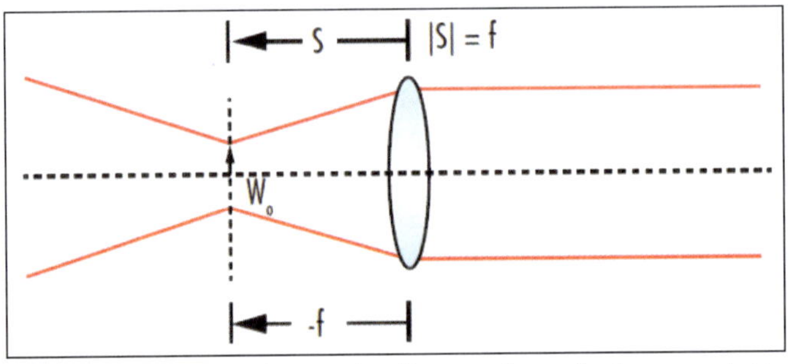

To collimate a Gaussian beam, the distance from the beam waist to the collimating lens should be equal to the lens' focal length.

References

- Fraunhofer-Diffraction, physics: scienceworld.wolfram.com, Retrieved 08 August, 2019
- Limits-of-resolution-the-rayleigh-criterion, physicstestbook2: opentextbc.ca, Retrieved 24 May, 2019
- Grating-physics: newport.com, Retrieved 25 July, 2019
- Diffraction-gratings: rp-photonics.com, Retrieved 16 April, 2019
- Fresnel-zone-plates, diffracting-optics, types-of-optics: x-ray-optics.de, Retrieved 18 May, 2019

4

Reflection and Refraction

Reflection is defined as the bouncing back of light from a surface. Refraction is defined as the bending of light as it passes from one medium to another. This chapter has been carefully written to provide an easy understanding of the concepts of reflection and refraction such as total internal reflection, image formation by lenses, etc. with respect to physical optics.

Reflection of Light

Spheres reflected in the floor and in each other.

Reflection occurs when a wave hits the interface between two dissimilar media, so that all of or at least part of the wave front returns into the medium from which it originated. Common examples are reflection of light, as shown in figure, as well as reflection of surface waves that may be observed in a pool of water, or sound waves reflected as echo from a wall.

Reflection of light may be specular or diffuse. The first occurs on a blank mirroring surface that retains the geometry of the beams of light. The second occurs on a rougher surface, not retaining the imaging geometry, only the energy.

Reflectance

Reflectance is the ratio of reflected power to incident power, generally expressed in decibels or percent. Most real objects have some mixture of diffuse and specular qualities, and surface reflectance is therefore often divided into diffuse and specular reflectance. In climatology, reflectivity is called albedo.

Diffuse Reflection

When light strikes a rough or granular surface, it bounces off in all directions due to the microscopic irregularities of the interface, as illustrated in figure. Thus, an image is not formed. This is called diffuse reflection. The exact form of the reflection depends on the structure of the surface.

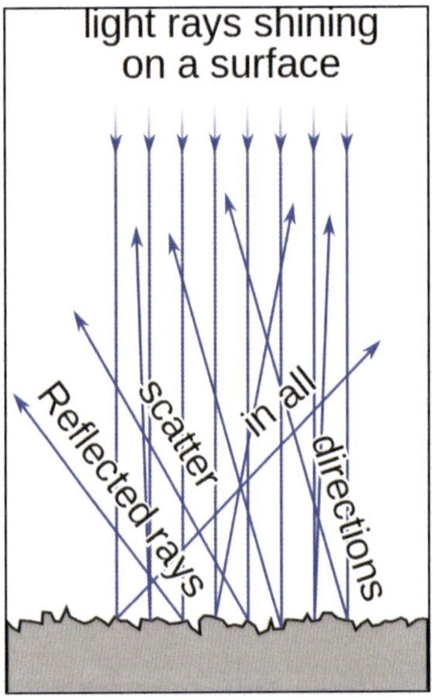

Diffuse reflection from a rough surface.

A common model for diffuse reflection is Lambertian reflectance, in which the light is reflected in accordance with Lambert's cosine law.

Lambert's cosine law says that the total radiant power observed from a "Lambertian" surface is directly proportional to the cosine of the angle θ between the observer's line of sight and the surface normal. The law is also known as the cosine emission law or Lambert's emission law.

When an area element is radiating as a result of being illuminated by an external source, the irradiance (energy/time/area) landing on that area element will be proportional to the cosine of the angle between the illuminating source and the normal. A Lambertian reflector will then reflect this light according to the same cosine law as a Lambertian emitter.

This means that although the radiance of the surface under diffuse reflection depends on the angle from the normal to the illuminating source, it will not depend on the angle from the normal to the observer.

An important consequence of Lambert's cosine law is that when an area element is viewed from any angle, it has the same radiance. Although the emitted power from an area element is reduced by the cosine of the emission angle, the observed size of the area element is also reduced by that same amount, so that while the area element appears smaller, its radiance is the same.

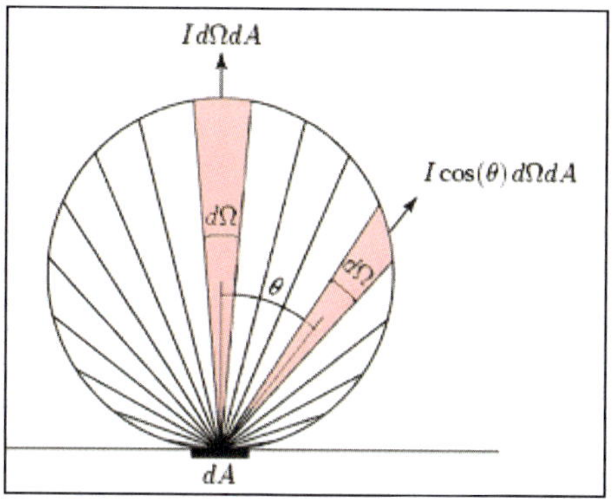

Emission rate (photons/sec) in a normal and off-normal direction. The number of photons/sec directed into any wedge is proportional to the area of the wedge.

Specular (Mirror-like) Reflection

We describe the directions of the incident ray hitting an optical surface and the ray being reflected from the surface in terms of the angle they make with the normal (perpendicular) to the surface at the point of incidence, as shown in figure. In specular reflection, the angle of incidence θ_i equals the angle of reflection θ_r.

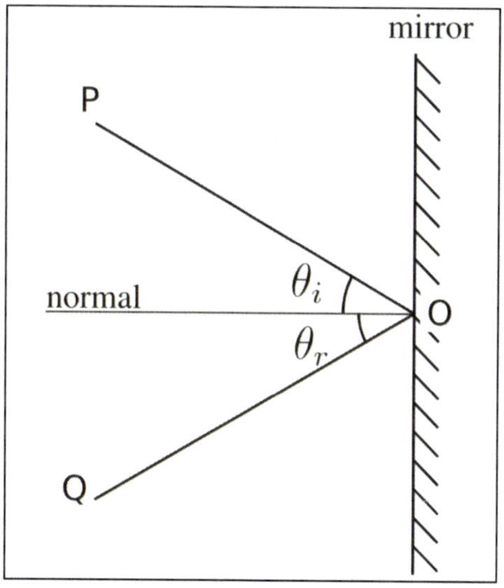

Reflection from a mirror.

A mirror is the most common example of specular light reflection. It usually consists of a glass sheet in front of a metallic coating where the reflection actually occurs. Reflection also occurs from the surface of transparent media, such as water or glass.

In fact, reflection of light may occur whenever light travels from a medium of a given refractive index into a medium with a different refractive index. In the most general case, a certain fraction of the light is reflected from the interface, and the remainder is refracted as it passes into the transparent medium.

When light reflects off a material denser than the external medium, it undergoes a 180° phase reversal. In contrast, a less dense, lower refractive index material will reflect light in phase. This is an important principle in the field of thin-film optics.

Specular reflection at a curved surface forms an image which may be magnified or demagnified. We will shortly return to the subject of curved mirrors and geometric optics.

Perfect Mirrors

A perfect mirror is a theoretical mirror that reflects light (and electromagnetic radiation in general) perfectly, and doesn't transmit it. Domestic mirrors are not perfect mirrors as they absorb a significant portion of the light which falls on them.

Dielectric mirrors are glass or other substrates on which one or more layers of dielectric material are deposited, to form an optical coating. A very complex dielectric mirror can reflect up to 99.999% of the light incident upon it, for a narrow range of wavelengths and angles. A simpler mirror may reflect 99.9% of the light, but may cover a broader range of wavelengths.

Image Formation by a Plane Mirror

As illustrated by figure, a plane mirror will form a virtual image point located exactly opposite the real image point, and as far behind the mirror as the object point is from the front of the mirror.

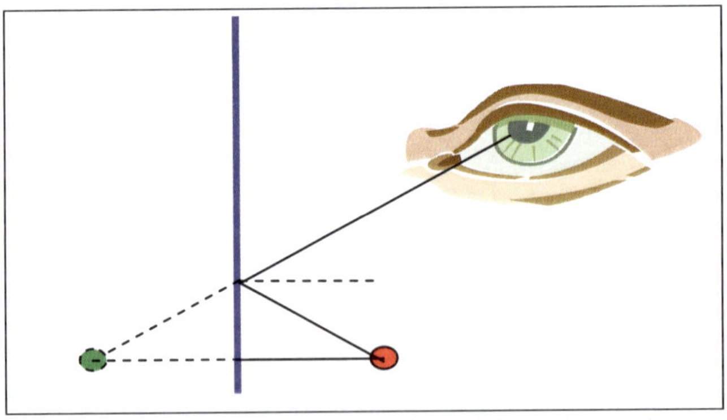

The location of the virtual image formed by a plane mirror.

Evidently, a "full size mirror" for your wardrobe doesn't have to more than half the height of your body, no matter where on your body your eyes are placed, and no matter how far away from the mirror you are standing. The image of an extended object seen in a plane mirror is exactly the same size as the object, but it is perceived to be located at a negative image distance behind the mirror, equal to the positive object distance from the mirror. And since the object and image sizes are the same, the lateral magnification is unity.

Retro Reflection

Retro reflectors are used both in a purely specular and a diffused mode. A simple retro reflector can be made by placing three ordinary mirrors mutually perpendicular to one another (a corner reflector). The image produced is the inverse of one produced by a single mirror. The principle is illustrated for two mirrors in figure. Applications also include simple radar reflectors for ships, and the Lunar Laser Range program based on four retro reflectors placed at the Apollo 11, 14, and 15 sites and the Lunakhod 2 rover. In such applications, it is imperative that the corner angle is exactly 90°. A deviation by an angle δ will cause the returned beam to deviate by 2δ from the incoming beam.

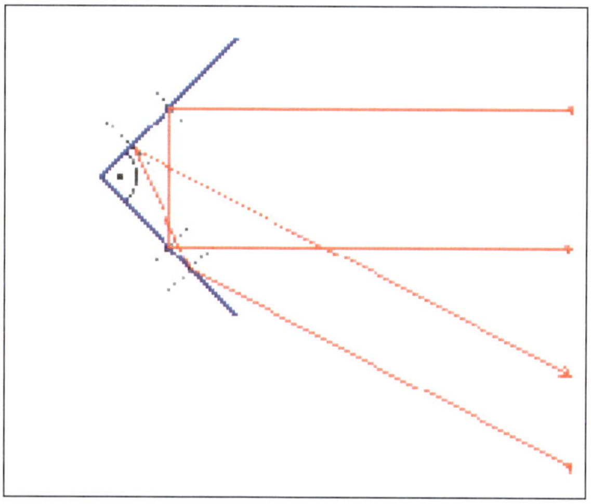

The principle of a retro reflector.

Surfaces may also be retroreflective. The structure of these surfaces is such that light is returned in the direction from which it came.

A surface can be made partially retroreflective by depositing a layer of tiny refractive spheres on it or by creating small pyramid like structures (cube corner reflection). In both cases internal reflection causes the light to be reflected back to where it originated. This is used to make traffic signs and automobile license plates reflect light mostly back in the direction from which it came. In this application a completely perfect retro reflection is not desired, since the light would then be directed back into the headlights of an oncoming car rather than to the driver's eyes.

The description of retro reflectors may illustrate an important property of all images formed by reflecting or refracting surfaces: An image formed by one surface can serve as the object for a second imaging surface. In the case of two orthogonal mirrors as shown in figure, mirror M_1 forms a virtual image P_1' of an object point P, and mirror M_2 forms another virtual image P_2'. Image P_1' serves as an object for mirror M_2, forming a virtual image P_3'. Likewise, image P_2' serves as an object for mirror M_1, forming the same virtual image P_3'.

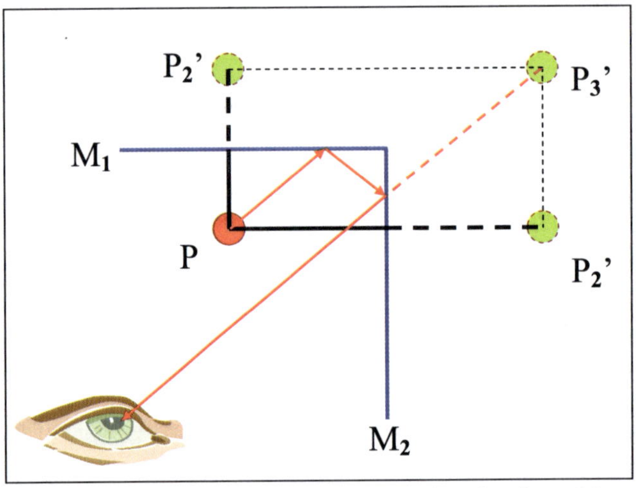

Virtual images P_1' and P_2' are formed by a single reflection of the object P in mirrors M_1 and M_2, respectively. The virtual image P_3' is formed by a reflection of the virtual image P_1' in mirror M_2 or image P_2' in mirror M_1.

This principle will be used extensively when locating the final image formed by several successive reflecting surfaces in a mirror-based telescope, or several refracting surfaces in a microscope or a refracting telescope.

Sign Rules for Image Formation

In the case of an image formed by a few plane mirrors, the geometry is quite simple, and we do not have to worry about the sign of object and image distances. However, we want to state some general sign rules that will be applicable to all imaging situations when both real and virtual images are formed in front of or behind curved surfaces:

- The object distance: When the object is on the same side of the reflecting or refracting surface as the incoming light, the object distance s is positive; otherwise, it is negative.
- The image distance: When the image is on the same side of the reflecting or refracting surface as the outgoing light, the image distance s' is positive; otherwise, it is negative.
- Radius of curvature: When the centre of curvature C is on the same side as the outgoing light, the radius of curvature is positive; otherwise, it is negative.

Image Formation by Curved Mirrors

Spherical Mirrors

Figure shows a spherical mirror with radius of curvature R. The concave side is facing the incident light originating from an object point P on the optical axis. A ray from P at an angle α to the axis is reflected by the mirror. The angle of incidence and reflection are both θ, and the reflected ray crosses the optical axis at an angle β. In fact, all rays from P will intersect the axis at the same point P', provided that the angle α is small.

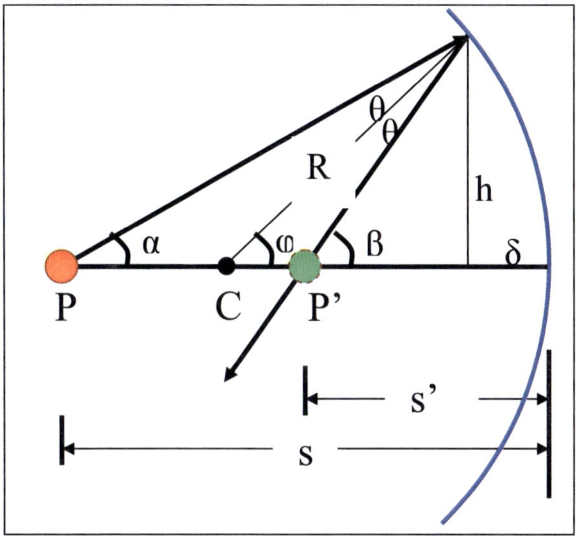

Finding the image point formed by a concave mirror.

We have the following relations: $\phi = \alpha + \theta$ and $\beta = \phi + \theta$, which implies that $\alpha + \beta = 2\phi$. The expressions for the tangents of α, β, and ϕ are simply $tg(\alpha) = h/(s-\delta)$, $tg(\beta) = h/(s'-\delta)$, $tg(\phi) = h/(R-\delta)$. If the angle α is small, so are β and φ. If α is so small that the ray is almost parallel to the optical axis (the paraxial approximation), the tangent of an angle is equal to the angle itself (given in radians), and δ may be neglected compared to s, s', and R. So for small angles we have the following approximations: $\alpha = h/s$, $\beta = h/s'$, $\phi = h/R$. Substituting this into φ = α + θ we get the general object-image relation for a spherical mirror:

$$\frac{1}{S} + \frac{1}{S'} = \frac{2}{R}$$

If the radius of curvature becomes infinite $(R = \infty)$, the mirror becomes plane, and the relation above reduces to $s = -s'$ for a plane reflecting surface.

If the object is very far from the mirror $(s \approx \infty)$, the incoming rays are parallel, and the image will be formed at a distance s' = R/2 from the mirror. This is the focal length, f.

A simple sketch may illustrate how the size y' and position s' of the real image of an extended object is determined, once the size y of the object, the distance s from the spherical mirror to the object and the radius R of curvature of the mirror is known.

We see that y/s = - y'/s'. The negative sign indicates that the image is inverted relative to the object. The magnification is m= y'/y = - s'/s. The size of the image is |y'| = ys'/s. Substituting s' from the general object-image relation, we get a more useful expression: |y'| = yf /(s-f), as the object distance s and the focal length f are often readily available.

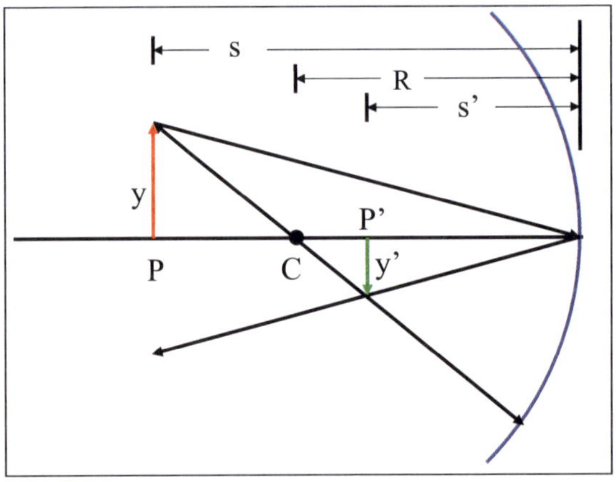

Determining the position, orientation and size of an image.

When the object is far from the mirror, the image of the object is smaller than the object, inverted, and real. If the object for all practical purposes is infinitely far away, the image is formed in the focal plane. As the object is moved closer to the mirror, the image moves farther from the mirror and increases in size. When the object is in the focal plane, the image is at infinity. If the object is inside the focal point, the image becomes larger than the object, erect, and virtual.

For a convex spherical mirror, R is negative. A virtual image is formed behind the mirror at a negative image distance. But using the sign rules, the same object-image relation is valid, and the expressions for the magnification and the size of the image are the same. However, if the incoming rays are parallel to the optical axis, the rays will not converge through a focal point. They will diverge as if they had come from a virtual focal point at negative focal length $f = R/2$ behind the mirror. In summary, the basic relationships for image formation by a spherical mirror are valid for both concave and convex mirrors, provided that the sign rules are used consistently.

Optical Aberrations Caused by Curved Mirrors

- Spherical aberration: A simple mirror with a spherical surface suffers from the optical imperfection called spherical aberration: Light striking nearer the periphery focuses closer to the mirror, while light striking near the center focuses farther away. While spherical mirrors are the most convenient to grind and polish, spherical aberration can be eliminated by grinding and polishing the surface to a paraboloid of revolution. Then parallel rays striking all parts of the mirror are reflected to the same focus.
- Coma: The principal disadvantage of a parabolic mirror is that it produces good images over only a relatively small field of view, that is, for light that strikes the mirror very nearly parallel to the optical axis. In a photograph of a field of stars, the star images near the center of the picture will appear as sharp, round dots, whereas those near the corners, which are formed by light coming in at an angle

to the optical axis, are distorted into tiny "tear drops" or "commas" pointing towards the center of the photograph. The shape of these images accounts for the name "coma" given to this type of aberration that distorts images formed by light that strikes the mirror off-axis.

- Astigmatism: Astigmatism occurs when rays of light in different planes do not focus at the same distance from the mirror. Such aberrations may be caused by mechanical distortions of large mirrors.
- Curvature of field: This is an aberration in which the image is sharp, but different parts of it are formed at different distances from the mirror, so that the whole image cannot be captured by a flat detector.
- Distortion: This is an aberration in which the image may be sharp, but its shape is distorted, e.g. if straight lines in the object plane are imaged as curved lines. Distortion may vary within the field of view, being most noticeable near the edges of the field of view.
- Vignetting: This is an aberration that causes a darkening of the image towards the corners of the field of view.

Reflecting Telescopes

A telescope is an instrument designed for the observation of remote objects. There are three main types of optical astronomical telescopes:

- The refracting (dioptric) telescope which uses an arrangement of lenses.
- The reflecting (catoptric) telescope which uses an arrangement of mirrors.
- The catadioptric telescope which uses a combination of mirrors and lenses.

The Italian monk Niccolo Zucchi is credited with making the first reflector in 1616, but he was unable to shape the concave mirror accurately and he lacked of means of viewing the image without blocking the mirror, so he gave up the idea. The first practical reflector design is due to James Gregory. In Optica Promota he described a reflector using two concave mirrors. A working example was not built until 10 years later by Robert Hooke. Sir Isaac Newton is often credited with constructing the first "practical" reflecting telescope after his own design in 1668. However, because of the difficulty of precisely producing the reflecting surface and maintaining a high and even reflectivity, the reflecting telescope did not become an important astronomical tool until a century later.

Telescopes increase the apparent angular size of distant objects, as well as their apparent brightness. If the telescope is used for direct viewing by the human eye, an eyepiece is used to view the image. And most, if not all eyepiece designs use an arrangement of lenses. However, in most professional applications the image is not viewed by the human eye, but is captured by photographic film or digital sensors, without the use of an eyepiece. In this configuration, telescopes are used as pure reflectors.

In a prime focus design in large observatory telescopes, the observer or image detector sits inside the telescope, at the focal point of the reflected light. In the past this would be the astronomer himself, but nowadays CCD cameras are used. Radio telescopes also often have a prime focus design.

Telescope Mountings

The telescope tube must be mounted so that it can point to any direction in the sky. This is achieved by using two orthogonal axes of motion.

- The simplest is an altitude-azimuth mount, where one axis points towards the zenith, and the instrument is pointed at a given altitude angle above the horizon, measured along a vertical circle, and a given azimuthal angle, measured eastward from the north point. Following a celestial object over time requires repeated recalculation of altitude and azimuth if this mount is used.
- In the equatorial mount, one axis is parallel to the earth's axis. This allows the telescope to be pointed directly in right ascension and declination, the coordinates in which astronomical positions are generally given. In addition, a simple clockwork about a single axis can compensate for the earth's rotation when observing celestial objects.

Optical Reflecting Telescope Designs

- The Newtonian usually has a paraboloid primary mirror but for small apertures, a spherical primary mirror is sufficient for high visual resolution. A flat diagonal secondary mirror mounted on a "spider" reflects the light to a focal plane at the side of the top of the telescope tube. Thus, the observer does not block the light entering the telescope tube. However, the observer may experience some awkward viewing positions.

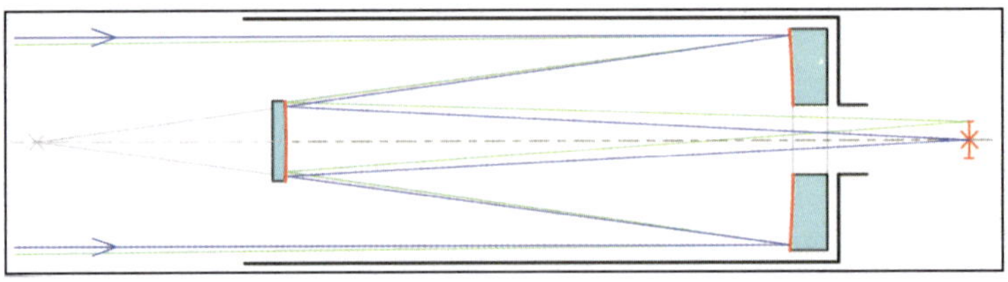

A "classic" Cassegrain telescope design.

- The "classic" Cassegrain has a parabolic primary mirror, and a convex hyperbolic secondary mirror that reflects the light back, either through a hole in the primary, or reflected out to the side of the tube by a third, flat mirror, mounted diagonally just in front of the primary (the Nasmyth-Cassegrain). Folding the optics makes this a compact design. On smaller telescopes, and camera lenses, the secondary is often mounted on a clear, optically flat, glass plate that closes the telescope tube. This eliminates the diffraction effects caused by the secondary supports in the Newtonian. It also protects the mirrors from dust.

- The Ritchey-Chrétien is a specialized Cassegrain reflector which has two hyperbolic mirrors (instead of a parabolic primary). It is free of coma and spherical aberration at a flat focal plane, making it well suited for wide field and photographic observations. Almost every professional reflector telescope in the world is of the Ritchey-Chrétien design.
- The Dall-Kirkham Cassegrain design uses a concave elliptical primary mirror and a convex spherical secondary. While this system is easier to grind than a classic Cassegrain or Ritchey-Chretien system, it does not correct for off-axis coma and field curvature so the image degrades quickly off-axis. Because this is less noticeable at longer focal ratios, Dall-Kirkhams are seldom faster than f/15.
- The Schiefspiegler uses tilted mirrors to avoid the secondary mirror casting a shadow on the primary. However, while eliminating diffraction patterns this leads to several other aberrations that must be corrected.
- The Maksutov telescope uses a full aperture corrector lens, a spherical primary, and a spherical secondary which is an integral part of the corrector lens.
- The Gregorian has a concave, not convex, secondary mirror and in this way achieves an upright image, useful for terrestrial observations.
- A Schmidt camera is an astronomical camera designed to provide wide fields of view with limited aberrations. It has a spherical primary mirror, and an aspherical correcting lens, located at the center of curvature of the primary mirror. The film or other detector is placed inside the camera, at the prime focus. The Schmidt camera is typically used as a survey instrument, for research programs in which a large amount of sky must be covered. There are several variations to the design:
    - Schmidt-Vaisala: A doubly-convex lens slightly in front of the film holder.
    - Baker-Schmidt: A convex secondary mirror, which reflected light back toward the primary. The photographic plate was then installed near the primary, facing the sky.
    - Baker-Nunn: Replaced the Baker-Schmidt camera's corrector plate with a small triplet corrector lens closer to the focus of the camera. A 20 inch Baker-Nunn camera installed near Oslo was used to track artificial satellites from the late 1950s.
    - Mersenne-Schmidt: Consists of a concave paraboloidal primary, a convex spherical secondary, and a concave spherical tertiary mirror.

- Schmidt-Newtonian: The addition of a flat secondary mirror at 45° to the optical axis of the Schmidt design creates a Schmidt-Newtonian telescope.
- Schmidt-Cassegrain: The addition of a convex secondary mirror to the Schmidt design directing light through a hole in the primary mirror creates a Schmidt-Cassegrain telescope. The last two designs are popular because they are compact and use simple spherical optics.

- The Nasmyth design is similar to the Cassegrain except no hole is drilled in the primary mirror; instead, a third mirror reflects the light to the side and out of the telescope tube, as shown in figure.
- Adding further optics to a Nasmyth style telescope that deliver the light (usually through the declination axis) to a fixed focus point that does not move as the telescope is reoriented gives you a Coudé focus. This design is often used on large observatory telescopes, as it allows the use of heavy observation equipment.

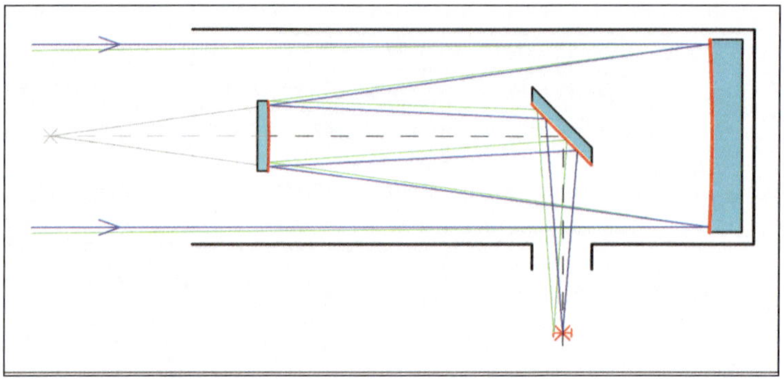

A Nasmyth Cassegrain design.

Non-optical Telescopes

Single-dish radio telescopes are often made of a conductive wire mesh whose openings are smaller than the reflected wavelength. The shape is often parabolic, and the detector is either placed in prime focus, or a secondary reflector is used.

Multi-element radio telescopes are constructed from pairs or groups of antennae to synthesize apertures that are similar in size to the separation between the telescopes. Large baselines are achieved by utilizing space-based Very Long Baseline Interferometry (VLBI) telescopes such as the Japanese HALCA (Highly Advanced Laboratory for Communications and Astronomy) VSOP (VLBI Space Observatory Program) satellite. The VLBI technique is also using radio telescopes on different continents simultaneously.

Image Formation by Concave Mirrors

There are two alternative methods of locating the image formed by a concave mirror. The first is purely graphical, and the second uses simple algebraic analysis.

The graphical method of locating the image produced by a concave mirror consists of drawing light-rays emanating from key points on the object, and finding where these rays are brought to a focus by the mirror. This task can be accomplished using just *four* simple rules:

1. An incident ray which is parallel to the principal axis is reflected through the focus *F* of the mirror.

2. An incident ray which passes through the focus *F* of the mirror is reflected parallel to the principal axis.

3. An incident ray which passes through the centre of curvature *C* of the mirror is reflected back along its own path (since it is normally incident on the mirror).

4. An incident ray which strikes the mirror at its vertex *V* is reflected such that its angle of incidence with respect to the principal axis is equal to its angle of reflection.

The validity of these rules in the paraxial approximation is fairly self-evident.

Consider an object *ST* which is placed a distance p from a concave spherical mirror, as shown in figure. For the sake of definiteness, let us suppose that the object distance *p* is greater than the focal length *f* of the mirror. Each point on the object is assumed to radiate light-rays in all directions. Consider four light-rays emanating from the tip *T* of the object which strike the mirror, as shown in the figure. The reflected rays are constructed using rules 1-4 above, and the rays are labelled accordingly. It can be seen that the reflected rays all come together at some point *T'*. Thus, T' is the image of *T* (*i.e.*, if we were to place a small projection screen at *T'* then we would see an image of the tip on the screen). As is easily demonstrated, rays emanating from other parts of the object are brought into focus in the vicinity of *T'* such that a complete image of the object is produced between *S'* and *T'* (obviously, point *S'* is the image of point *S*). This image could be viewed by projecting it onto a screen placed between points *S'* and *T'*. Such an image is termed a *real image*. Note that the image *S'T'* would also be directly visible to an observer looking straight at the mirror from a distance greater than the image distance *q* (since the observer›s eyes could not tell that the light-rays diverging from the image were in anyway different from those which would emanate from a real object). According to the figure, the image is *inverted* with respect to the object, and is also magnified.

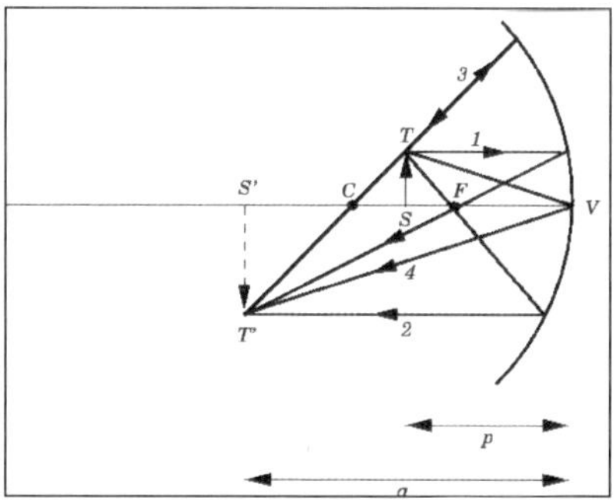

Formation of a real image by a concave mirror.

Figure shows what happens when the object distance p is less than the focal length f. In this case, the image appears to an observer looking straight at the mirror to be located behind the mirror. For instance, rays emanating from the tip *T* of the object appear, after reflection from the mirror, to come from a point *T'* which is behind the mirror. Note that only two rays are used to locate *T'*, for the sake of clarity. In fact, two is the minimum number of rays needed to locate a point image. Of course, the image behind the mirror cannot be viewed by projecting it onto a screen, because there are no real light-rays behind the mirror. This type of image is termed a virtual image. The characteristic difference between a real image and a virtual image is that, immediately after reflection from the mirror, light-rays emitted by the object converge on a real image, but diverge from a virtual image. According to figure, the image is upright with respect to the object, and is also magnified.

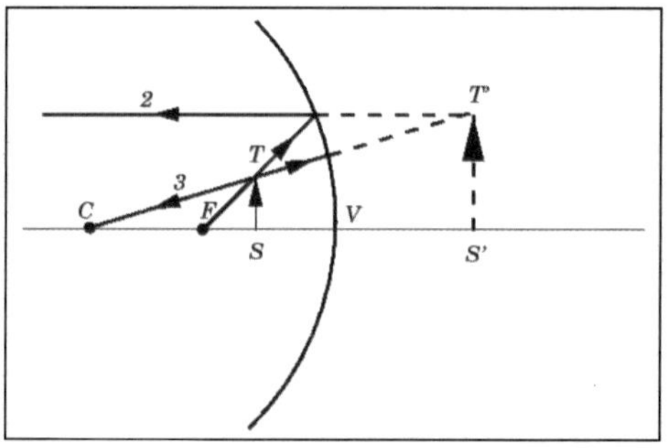

Formation of a virtual image by a concave mirror.

The graphical method described above is fine for developing an intuitive understanding of image formation by concave mirrors, or for checking a calculation, but is a bit too cumbersome for everyday use. The analytic method described below is far more flexible.

Consider an object *ST* placed a distance p in front of a concave mirror of radius of curvature R. In order to find the image *S'T'* produced by the mirror, we draw two rays from *T* to the mirror. The first, labelled 1, travels from *T* to the vertex *V* and is reflected such that its angle of incidence θ equals its angle of reflection. The second ray, labelled 2, passes through the centre of curvature *C* of the mirror, strikes the mirror at point *B*, and is reflected back along its own path. The two rays meet at point *T'*. Thus, *S'T'* is the image of *ST*, since point *S'* must lie on the principal axis.

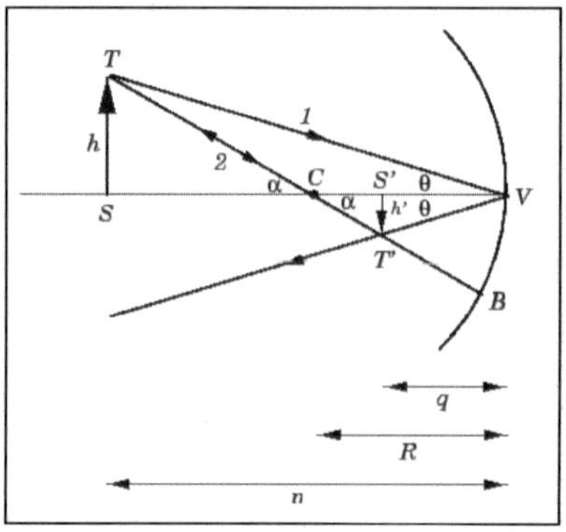

Image formation by a concave mirror.

In the triangle *STV*, we have $\tan\theta = h/p$, and in the triangle $S'T'V$ we have $\tan\theta = -h'/q$, where *p* is the object distance, and *q* is the image distance. Here, *h* is the height of the object, and *h'* is the height of the image. By convention, *h'* is a negative number, since the image is inverted (if the image were upright then *h'* would be a positive number). It follows that,

$$\tan\theta = \frac{h}{p} = \frac{-h'}{q}.$$

Thus, the magnification M of the image with respect to the object is given by,

$$M = \frac{h'}{h} = -\frac{q}{p}.$$

By convention, M is negative if the image is inverted with respect to the object, and positive if the image is upright. It is clear that the magnification of the image is just determined by the ratio of the image and object distances from the vertex.

From triangles *STC* and *S'T'C*, we have $\tan\alpha = h/(p-R)$ and $\tan\alpha = -h'/(R-q)$, respectively. These expressions yield,

$$\tan\alpha = \frac{h}{p-R} = -\frac{h'}{R-q}.$$

Equations $M = \frac{h'}{h} = -\frac{q}{p}$ and $\tan\alpha = \frac{h}{p-R} = -\frac{h'}{R-q}$ can be combined to give,

$$\frac{-h'}{h} = \frac{R-q}{p-R} = \frac{q}{p},$$

Which easily reduces to,

$$\frac{1}{p} + \frac{1}{q} = \frac{2}{R}.$$

This expression relates the object distance, the image distance, and the radius of curvature of the mirror.

For an object which is very far away from the mirror (*i.e.*, $p \to \infty$), so that light-rays from the object are parallel to the principal axis, we expect the image to form at the focal point *F* of the mirror. Thus, in this case, $q = f$, where *f* is the focal length of the mirror, and equation $\frac{1}{p} + \frac{1}{q} = \frac{2}{R}$ reduces to,

$$0 + \frac{1}{f} = \frac{2}{R}.$$

The above expression yields,

$$f = \frac{R}{2}.$$

In other words, in the paraxial approximation, the focal length of a concave spherical mirror is half of its radius of curvature. Equations $\frac{1}{p} + \frac{1}{q} = \frac{2}{R}$ and $f = \frac{R}{2}$ can be combined to give,

$$\frac{1}{p} + \frac{1}{q} = \frac{1}{f}.$$

The above expression was derived for the case of a real image. However, as is easily demonstrated, it also applies to virtual images provided that the following sign convention is adopted. For real images, which always form *in front* of the mirror, the image distance q is *positive*. For virtual images, which always form *behind* the mirror, the image distance q is negative. It immediately follows, from equation $M = \frac{h'}{h} = -\frac{q}{p}$, that real images are always inverted, and virtual images are always upright. Table shows how the location and character of the image formed in a concave spherical mirror depend on the location of the object, according to equations $M = \frac{h'}{h} = -\frac{q}{p}$ and $\frac{1}{p} + \frac{1}{q} = \frac{1}{f}$. It is clear that the *modus operandi* of a shaving mirror, or a makeup mirror, is to place the object (*i.e.*, a face) between the mirror and the focus of the mirror. The image is upright, (apparently) located behind the mirror, and magnified.

Table: Rules for image formation by concave mirrors.

Position of Object	Position of Image	Character of Image
At ∞	At F	Real, zero size
Between ∞ and C	Between F and C	Real, inverted, diminished
At C	At C	Real, inverted, same size
Between C and F	Between C and ∞	Real, inverted, magnified
At F	At ∞	
Between F and V	From −∞ to V	Virtual, upright, magnified
At V	At V	Virtual, upright, same size

Image Formation by Convex Mirrors

The definitions of the principal axis, centre of curvature *C*, radius of curvature *R*, and the vertex *V*, of a convex mirror are analogous to the corresponding definitions for a concave mirror. When parallel light-rays strike a convex mirror they are reflected such that they appear to emanate from a single point *F* located behind the mirror, as shown in figure. This point is called the virtual focus of the mirror. The focal length *f* of the mirror is simply the distance between *V* and *F*. As is easily demonstrated, in the paraxial approximation, the focal length of a convex mirror is half of its radius of curvature.

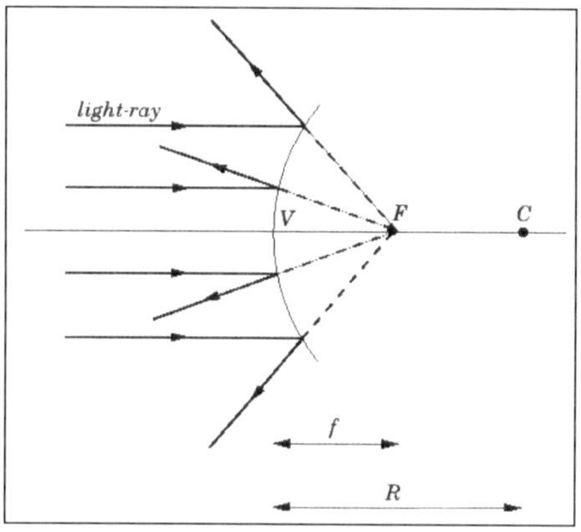

The virtual focus of a convex mirror.

There are, again, two alternative methods of locating the image formed by a convex mirror. The first is graphical, and the second analytical.

According to the graphical method, the image produced by a convex mirror can always be located by drawing a ray diagram according to four simple rules:

1. An incident ray which is parallel to the principal axis is reflected as if it came from the virtual focus *F* of the mirror.
2. An incident ray which is directed towards the virtual focus *F* of the mirror is reflected parallel to the principal axis.
3. An incident ray which is directed towards the centre of curvature *C* of the mirror is reflected back along its own path (since it is normally incident on the mirror).
4. An incident ray which strikes the mirror at its vertex *V* is reflected such that its angle of incidence with respect to the principal axis is equal to its angle of reflection.

The validity of these rules in the paraxial approximation is, again, fairly self-evident.

In the example shown in figure, two rays are used to locate the image $S'T'$ of an object *ST* placed in front of the mirror. It can be seen that the image is virtual, upright, and diminished.

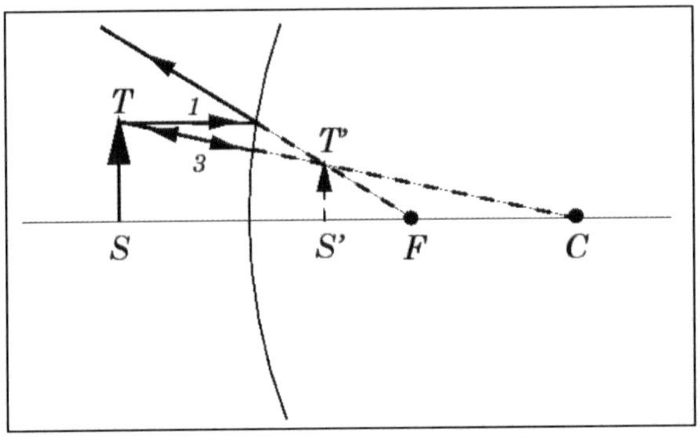

Image formation by a convex mirror.

As is easily demonstrated, application of the analytical method to image formation by convex mirrors again yields equation $M = \frac{h'}{h} = -\frac{q}{p}$ for the magnification of the image, and equation $\frac{1}{p} + \frac{1}{q} = \frac{1}{f}$ for the location of the image, provided that we adopt the following sign convention. The focal length f of a convex mirror is redefined to be *minus* the distance between points *V* and *F*. In other words, the focal length of a concave mirror, with a real focus, is always positive, and the focal length of a convex mirror, with a virtual focus, is always negative. Table shows how the location and character of the image formed in a convex spherical mirror depend on the location of the object, according to equations $M = \frac{h'}{h} = -\frac{q}{p}$ and $\tan\alpha = \frac{h}{p-R} = -\frac{h'}{R-q}$ (with $f < 0$).

Table: Rules for image formation by convex mirrors.

Position of object	Position of image	Character of image
At ∞	At F	Virtual, zero size
Between ∞ and V	Between F and V	Virtual, upright, diminished
At V	At V	Virtual, upright, same size

The formation of an image by a spherical mirror involves the *crossing* of light-rays emitted by the object and reflected off the mirror. If the light-rays actually cross in front of the mirror then the image is real. If the light-rays do not actually cross, but appear to cross when projected backwards behind the mirror, then the image is virtual. A real image can be projected onto a screen, a virtual image cannot. However, both types of images can be seen by an observer, and photographed by a camera. The magnification of the image is specified by equation $M = \frac{h'}{h} = -\frac{q}{p}$, and the location of the image is determined by equation $\frac{1}{p} + \frac{1}{q} = \frac{1}{f}$. These two formulae can be used to characterize both real and virtual images formed by either concave or convex mirrors, provided that the following sign conventions are observed:

1. The height h' of the image is positive if the image is upright, with respect to the object, and negative if the image is inverted.
2. The magnification M of the image is positive if the image is upright, with respect to the object, and negative if the image is inverted.
3. The image distance q is positive if the image is real, and, therefore, located in front of the mirror, and negative if the image is virtual, and, therefore, located behind the mirror.
4. The focal length f of the mirror is positive if the mirror is concave, so that the focal point F is located in front of the mirror, and negative if the mirror is convex, so that the focal point F is located behind the mirror.

Note that the front side of the mirror is defined to be the side from which the light is incident.

Refraction of Light

The reason why a glass lens may focus light is that light travels more slowly within the glass than in the medium surrounding it. The result is that the light rays are bent. This is described by the refraction index of the glass lens, which is equivalent to the ratio of the speed of light in air and glass.

We may illustrate this phenomenon by a simple example. Consider a ray of light passing through the air and entering a plane parallel slab of glass as shown in figure. Then Snell's law gives the relation between the angle of incidence (θ_1), the angle of refraction (θ_2), the refraction index of the glass slab (n_2) and the surrounding medium (n_1):

$$n_1 \sin\theta_1 = n_2 \sin\theta_2.$$

The angle of incidence and the refraction angle are the angles between the light beams and the normal to the surface at the point where the beam crosses the interface.

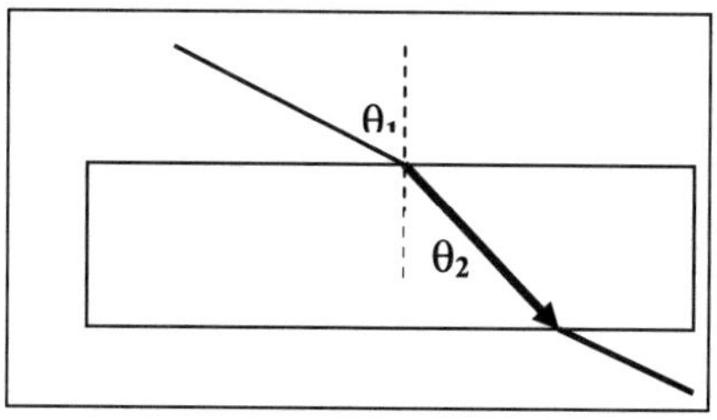

A light beam is bent as it enters a slab of glass – refraction.

A Simple Derivation of Snell's Law

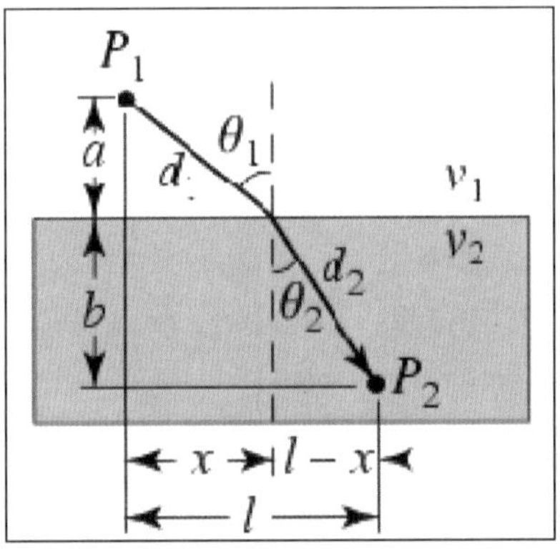

If we consider a beam of light passing at an angle from air into a plane-parallel slab of glass as illustrated in figure, then Snell's law gives a relationship between the angle of incidence, (θ_1), the refraction angle (θ_2), the refraction index of the glass slab (n_2), and the refraction index of the surrounding air (n_1):

$$n_1 \sin \theta_1 = n_2 \sin \theta_2.$$

The angles of incidence and refraction are the angles between the light beams and the surface normal at the entry point.

Now we will give a simple proof of Snell's law using the sketch to the right, together with Fermat's principle:

"A light ray will follow the path between two points that takes the shortest time."

The velocity of light in the two media is given in the sketch. And as always: distance = velocity × time.

Consider the time it takes to get from P_1 to P_2 in the sketch. This time is given by,

$$t = \frac{d_1}{v_1} + \frac{d_2}{v_2} = \frac{\sqrt{a^2 + x^2}}{v_1} + \frac{\sqrt{(l - x)^2 + b^2}}{v_2}$$

v_1 and v_2 are the velocities in the two media, and everything else is given in the sketch.

We want to adhere to Fermat's principle and obtain the least time consuming path. Finding the derivative of time with regard to the x-coordinate of the entry point, and setting this derivative to zero, we find the x-value that gives a minimum value of the time.

$$\frac{\partial t}{\partial x} = \frac{\frac{1}{2}\left(a^2 + x^2\right)^{-1/2} 2x}{v_1} + \frac{\frac{1}{2}\left[(l - x)^2 + b^2\right]^{-1/2} 2(l - x)(-1)}{v_2} = 0$$

$$\frac{x\left(a^2+x^2\right)^{-1/2}}{v_1}=\frac{(l-x)\left[(l-x)^2+b^2\right]^{-1/2}}{v_2}$$

Rearranging a bit, we see that,

$$\frac{1}{v_1}\frac{x}{\sqrt{a^2+x^2}}=\frac{1}{v_2}\frac{1-x}{\sqrt{(l-x)^2+b^2}}\Rightarrow\frac{1}{v_1\sqrt{\left(\frac{a}{x}\right)^2+1}}=\frac{1}{v_2\sqrt{\left(\frac{b}{l-x}\right)^2+1}}$$

From the sketch we see that tg θ_1 = x/a, while tg θ_2 = $(l-x)$/b, which gives us,

$$\frac{1}{v_1}\frac{x}{\sqrt{\cot g^2\theta_1+1}}=\frac{1}{v_2}=\frac{l-x}{\sqrt{\cot g^2\theta_2+1}}$$

And since,

$$\sin\theta=\frac{1}{\sqrt{\cot g^2\theta+1}}$$

this is equivalent to,

$$\frac{\sin\theta_1}{v_1}=\frac{\sin\theta_2}{v_2}$$

We then use the fact that the index of refraction is defined as $n_i \equiv c/v_i$, where *c* is the velocity of light, and arrive at Snell's law (for planar waves):

$$n_1\sin\theta_1=n_2\sin\theta_2$$

Which was discovered 400 years ago by the Dutch Willebrord Snell.

Usually, the index of refraction is given relative to the surrounding medium, and we get the expression,

$$\sin\alpha/\sin\beta=\mathrm{n}$$

where α is the angle of incidence and β is the angle of refraction.

The Refractive Index

Many materials have a well-characterized refractive index, but these indices depend strongly upon the wavelength of light. Therefore, any numeric value for the index is meaningless unless the associated wavelength is specified.

There are also weaker dependencies on temperature, pressure, and stress, as well on the precise material compositions (including the presence of impurities and dopants).

However, these variations are usually at the percent level or less. Thus, it is especially important to cite the source for an index measurement if high precision is claimed.

In general, an index of refraction is a complex number with both a real and imaginary part, where the latter indicates the strength of the relative absorption loss at a particular wavelength. The imaginary part is sometimes called the extinction coefficient k.

Table: The refractive index of some materials.

Material	n at λ = 589.3 nm
Vacuum	1 (exactly)
Air @ STP	1.0002926
Water (20°C)	1.333
Acrylic glass	1.490 - 1.492
Crown glass (pure)	1.50 - 1.54
Flint glass (pure)	1.60 - 1.62
Crown glass (impure)	1.485 - 1.755
Flint glass (impure)	1.523 - 1.925
Diamond	2.419
Gallium(III) arsenide	3.927

The Sellmeier Equation

Deduced in 1871 by W. Sellmeier, this equation models dispersion in a refractive medium, and is an empirical relationship between the refractive index *n* and the wavelength λ for a particular transparent medium. The usual form of the Sellmeier equation for glasses is:

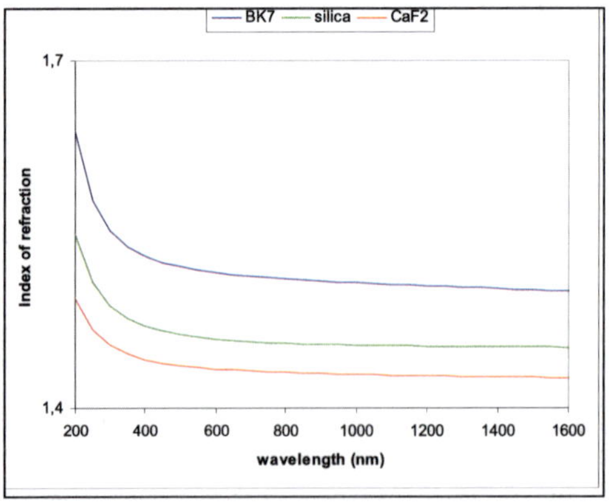

The refractive index of BK7, silica, and CaF_2 as a function of wavelength.

$$n(\lambda) = \left[1 + \sum_i \frac{B_i \lambda^2}{\lambda^2 - C_i}\right]^2$$

The coefficients $B_{1,2,3}$ and $C_{1,2,3}$ are usually quoted for λ in micrometers. Note that this λ is the vacuum wavelength; not that in the material itself, which is $\lambda/n(\lambda)$.

The Fresnel Equations

When light moves from a medium with refractive index n_1 into a second medium with refractive index n_2, both reflection and refraction of the light may occur, as shown in figure. The relationship between the angles that the incident, reflected and refracted rays make to the normal of the interface (given as θ_i, θ_r and θ_t, in figure), is given by the law of reflection and Snell's law, respectively.

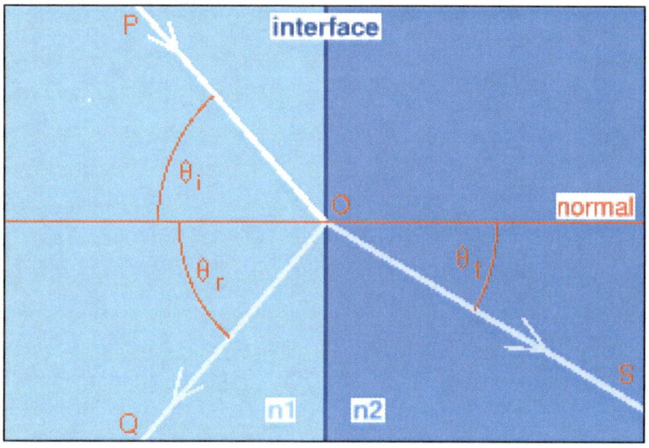

Reflection and refraction when passing from one medium into another.

Based on the assumption that the two materials are both non-magnetic, the Fresnel equations may be used to calculate the reflection coefficient R, i.e. the fraction of the intensity of incident light that is reflected from the interface. The reflection of light that these equations predict is known as Fresnel reflection.

The calculations of *R* depend on the polarization of the incident ray. If the light is polarized with the electric field of the light perpendicular to the plane of the diagram above (*s*-polarized), the reflection coefficient is given by:

$$R_S = \left[\frac{\sin(\theta_t - \theta_i)}{\sin(\theta_t - \theta_i)}\right]^2 = \left[\frac{n_1 \cos(\theta_i) - n_2 \cos(\theta_t)}{n_1 \cos(\theta_i) + n_2 \cos(\theta_t)}\right]^2$$

since $\sin(x \pm y) = \sin(x)\cos(y) \pm \cos(x)\sin(y)$, and $n_1 \sin(\theta_i) = n_2 \sin(\theta_t)$.

If the incident light is polarized in the plane of the diagram (p-polarized), the *R* is given by:

$$R_P \left[\frac{tg(\theta_t - \theta_i)}{tg(\theta_t + \theta_i)}\right]^2$$

The transmission coefficient in each case is given by $T_s = 1 - R_s$ and $T_p = 1 - R_p$. If the incident light is unpolarized (containing an equal mix of s- and p-polarizations), the reflection coefficient is $R = (R_s + R_p)/2$.

At one particular angle for a given n_1 and n_2, the value of R_p goes to zero and a ppolarized incident ray is purely refracted. This angle is known as Brewster's angle, and is around 56° for a glass medium in air or vacuum.

When moving from a more dense medium into a less dense one (i.e. $n_1 > n_2$), above an incidence angle known as the critical angle, all light is reflected and $R_s = R_p = 1$, as illustrated in figure. This phenomenon is known as total internal reflection. The critical angle is approximately 41° for glass in air.

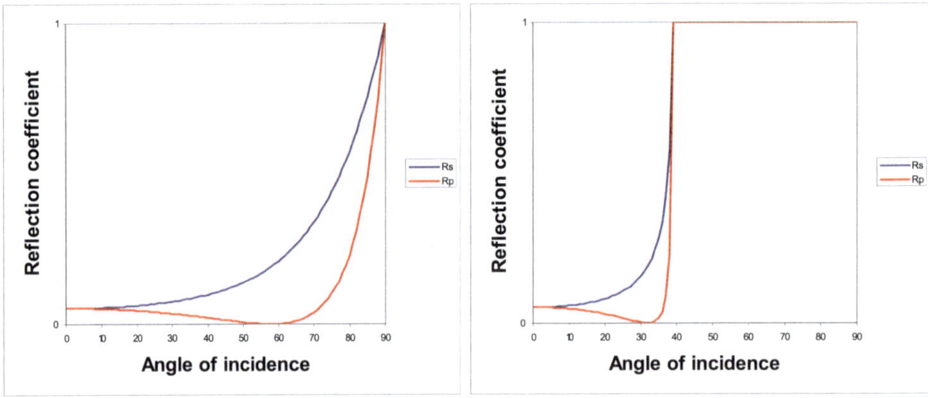

The reflection coefficients from air to glass (left) and glass to air (right).

When the light is at near-normal incidence to the interface $(\theta_i \approx \theta_t \approx 0)$, the reflection and transmission coefficient are given by:

$$R = R_S = R_P = \left[\frac{n_1 - n_2}{n_1 + n_2}\right]^2$$

$$T = T_S = T_P = 1 - R = \frac{4n_1 n_2}{(n_1 + n_2)^2}$$

For common glass, the reflection coefficient is about 4%. Note that reflection by a window is from the front side as well as the back side, and that some of the light bounces back and forth a number of times between the two sides. The combined reflection coefficient for this case is 2R/(1 + R), when interference can be neglected.

In reality, when light makes multiple reflections between two parallel surfaces, the multiple beams of light generally interfere with one another, and the surfaces act as a Fabry-Perot interferometer. This effect is responsible for the colors seen in oil films on water, and it is utilized to make antireflective optical coatings and optical filters.

Image Formation by Concave Lens

- Object is at infinity.

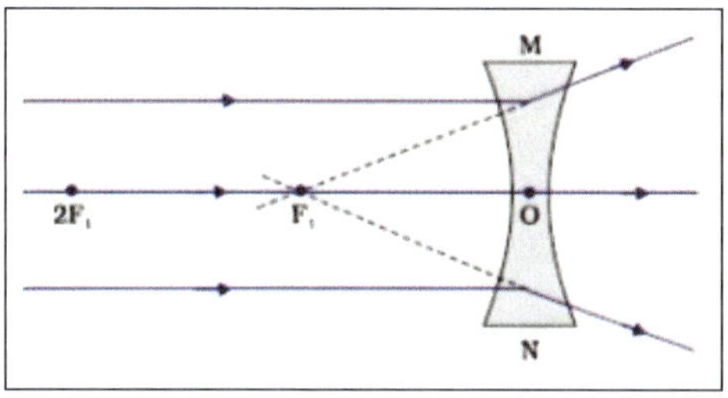

- Image is formed at F.
- It is highly diminished, virtual and erect.
- Object is between infinity and optical centre of lens.

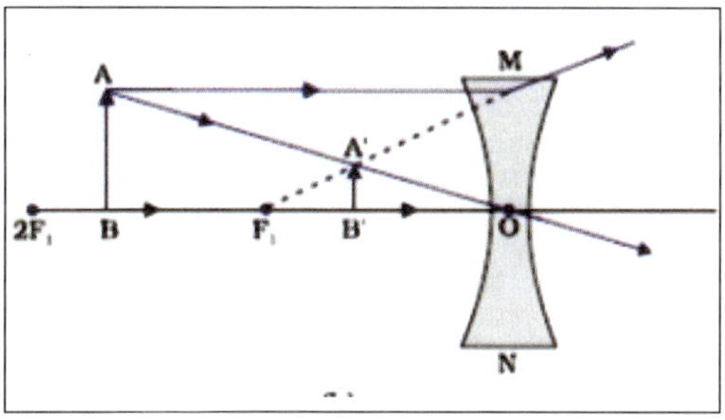

- Image is formed between focus and optical centre of the lens.
- It is virtual and erect.
- Sign conventions for spherical lenses
- All distances are measured from optical centre of a lens.
- Distances measured in direction of propagation of incident ray of light are taken as positive, while distances measured in a direction opposite to the direction of incident ray of incident are taken as negative.
- Upward heights wrt principal axis are taken as positive, downward height wrt to ray of light are taken as negative.

Image Formation by Convex Lens

The Object is Located Beyond 2F

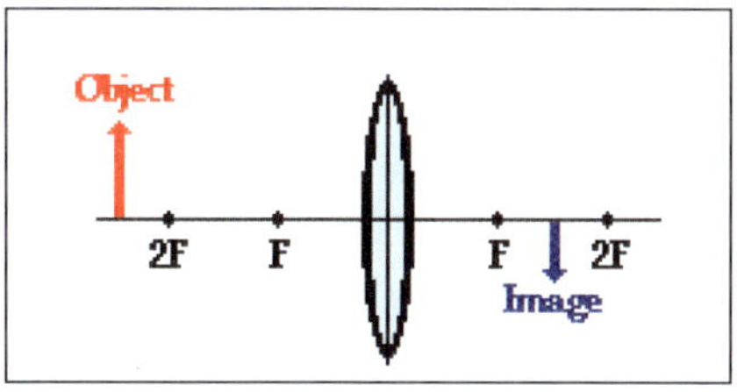

When the object is located at a location beyond the 2F point, the image will always be located somewhere in between the 2F point and the focal point (F) on the other side of the lens. Regardless of exactly where the object is located, the image will be located in this specified region. In this case, the image will be an inverted image. That is to say, if the object is right side up, then the image is upside down. In this case, the image is reduced in size; in other words, the image dimensions are smaller than the object dimensions. If the object is a six-foot tall person, then the image is less than six feet tall. The term magnification was introduced; the magnification is the ratio of the height of the object to the height of the image. In this case, the magnification is a number with an absolute value less than 1. Finally, the image is a real image. Light rays actually converge at the image location. If a sheet of paper were placed at the image location, the actual replica or likeness of the object would appear projected upon the sheet of paper.

The Object is Located at 2F

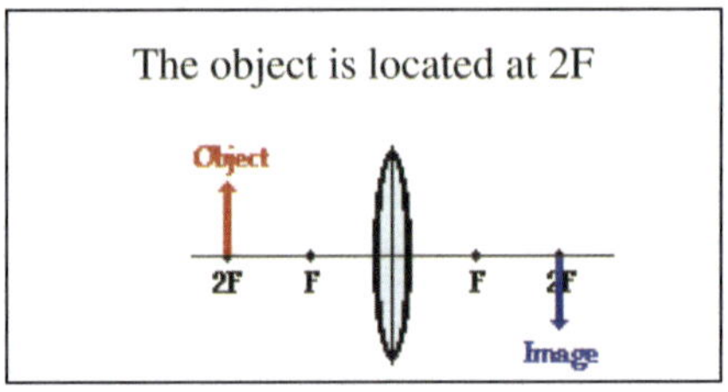

When the object is located at the 2F point, the image will also be located at the 2F point on the other side of the lens. In this case, the image will be inverted (i.e., a right side up object results in an upside-down image). The image dimensions are equal to the object dimensions. A six-foot tall person would have an image that is six feet tall; the absolute value of the magnification is exactly 1. Finally, the image is a real image. Light rays actually converge at the image location. As such, the image of the object could be projected upon a sheet of paper.

The Object is Located Between 2F and F

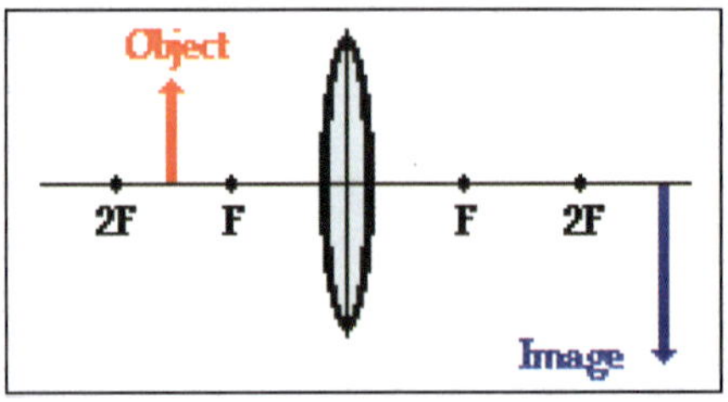

When the object is located *in front of* the 2F point, the image will be located *beyond* the 2F point on the other side of the lens. Regardless of exactly where the object is located between 2F and F, the image will be located in the specified region. In this case, the image will be inverted (i.e., a right side up object results in an upside-down image). The image dimensions are larger than the object dimensions. A six-foot tall person would have an image that is larger than six feet tall. The absolute value of the magnification is greater than 1. Finally, the image is a real image. Light rays actually converge at the image location. As such, the image of the object could be projected upon a sheet of paper.

The Object is Located at F

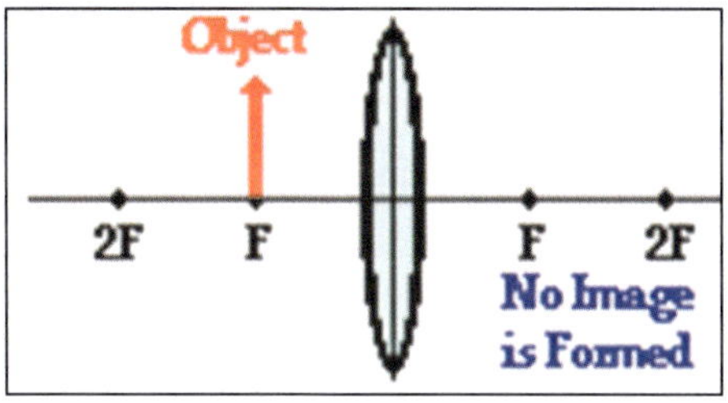

When the object is located at the focal point, no image is formed. The refracted rays neither converge nor diverge. After refracting, the light rays are traveling parallel to each other and cannot produce an image.

The Object is Located in Front of F

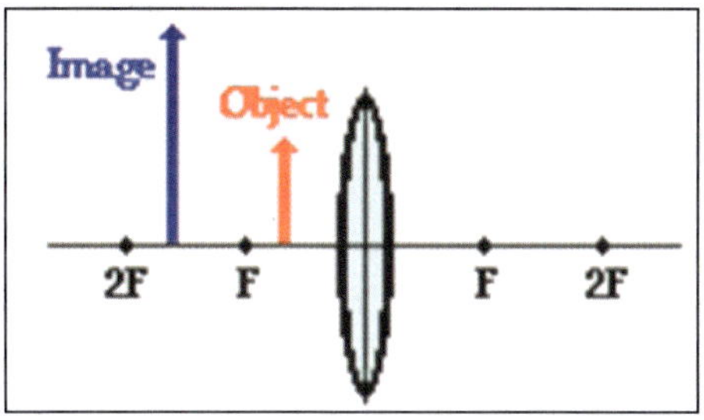

When the object is located at a location in front of the focal point, the image will always be located somewhere on the same side of the lens as the object. Regardless of exactly where in front of F the object is located, the image will always be located on the object's side of the lens and somewhere further from the lens. The image is located behind the object. In this case, the image will be an upright image. That is to say, if the object is right side up, then the image will also be right side up. In this case, the image is enlarged; in other words, the image dimensions are greater than the object dimensions. A six-foot tall person would have an image that is larger than six feet tall. The magnification is greater than 1. Finally, the image is a virtual image. Light rays diverge upon refraction; for this reason, the image location can only be found by extending the refracted rays backwards on the object's side the lens. The point of their intersection is the virtual image location. It would appear to any observer as though light from the object were diverging from this location. Any attempt to project such an image upon a sheet of paper would fail since light does not actually pass through the image location.

It might be noted from the above descriptions that there is a relationship between the object distance and object size and the image distance and image size. Starting from a large value, as the object distance decreases (i.e., the object is moved closer to the lens), the image distance increases; meanwhile, the image height increases. At the 2F point, the object distance equals the image distance and the object height equals the image height. As the object distance approaches one focal length, the image distance and image height approaches infinity. Finally, when the object distance is equal to exactly one focal length, there is no image. Then altering the object distance to values less than one focal length produces images that are upright, virtual and located on the same side of the lens as the object. Finally, if the object distance approaches 0, the image distance approaches 0 and the image height ultimately becomes equal to the object height. These patterns are depicted in the diagram below. Eight different object locations are drawn in red and labeled with a number; the corresponding image locations are drawn in blue and labeled with the identical number.

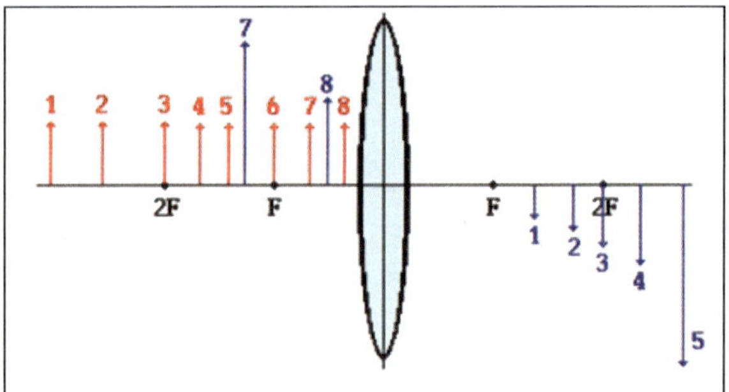

Each of the numbered objects (except #6) has an image with the corresponding number; its relative location, size, and orientation are shown.

Dispersion of Light

Dispersion of Light can be defined as the splitting of white light when it passes through a glass prism into its constituent spectrum of colors (i.e. violet, indigo, blue, green, yellow, orange and red). Dispersion figuratively means 'distribution' and hence that's exactly what is happening in the picture above. The white light splits into its constituent colors at various frequencies and various angles.

Prism

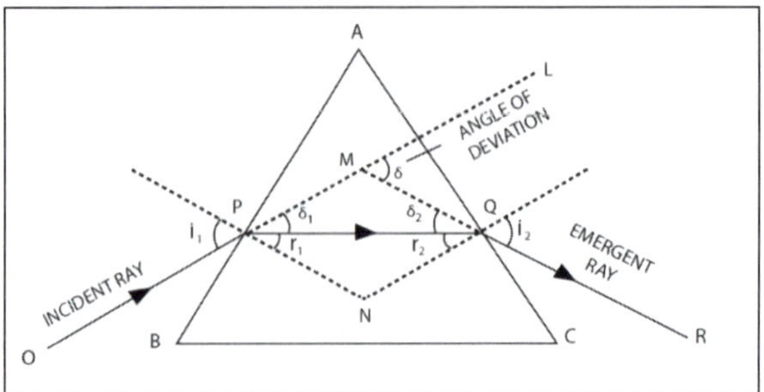

A prism is a transparent refracting device bounded by five plane surfaces inclined at some angles. It bends the light two times and emergent ray is at an angle to the incident ray.

Angle of Deviation

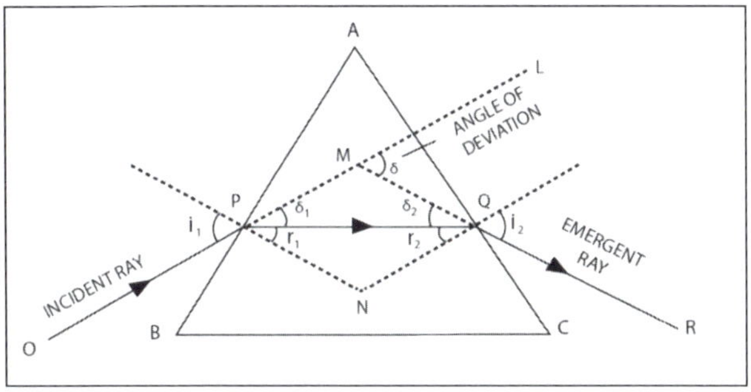

Angle of deviation (δ) is the angle between emergent ray and incident ray.

For a single refracting surface, $\delta = |i - r|$

For a prism, $\delta = (i_1 + i_2) - (r_1 + r_2)$,

$$\delta = i_1 + i_2 - A$$

Where A is the angle of the prism. For angle of minimum deviation, δ is minimum and $i_1 = i_2 = i$

$$\delta_{min} = 2i - A$$

For small A, $\delta = (\mu - 1)A$

Dispersion of White Light by a Prism

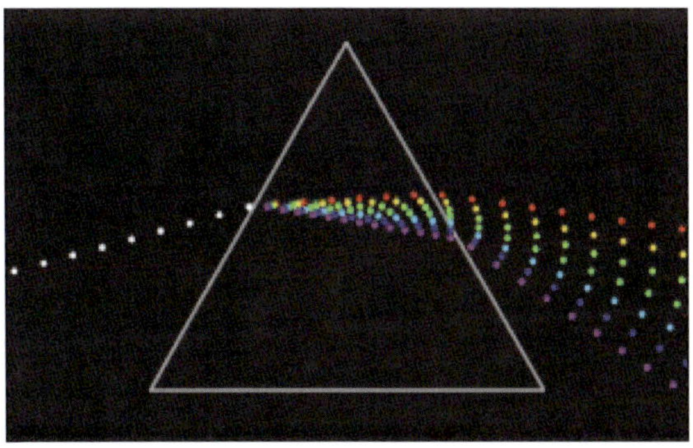

Refraction is the bending in the path of the light when it travels from one medium to the another. The degree at which refraction will occur depends on the wavelength of the light. Each light wave has a different wavelength and will therefore deviate differently.

White light is composed of light of different wavelengths (colors) i.e. violet, indigo, blue, green, yellow and red. Red has the highest wavelength and violet the lowest.

Wavelength is inversely proportional to the deviation in the path of the light. Red light suffers the least amount of deviation and violet the most. When a white light is made to pass through a prism, formation of a spectrum of seven colors occurs showing white light is a combination of seven separate colors.

Prism only acts as a medium for the dispersion of light made of the seven colors. Refraction occurs when the light falls on the prism. The wavelength and frequency of these deviated colors is different, they deviate differently at different angles due to the velocity difference of the prism. The color red therefore deviates the least since it has maximum wavelength and color violet deviates the most since it has the least wavelength.

Refraction of Light through a Glass Prism

A glass prism has 3 rectangular lateral surfaces and 2 triangular bases, all inclined at an angle. This angle is called the 'Angle of the prism'. Let's take a triangular prism with a ray of light entering it.

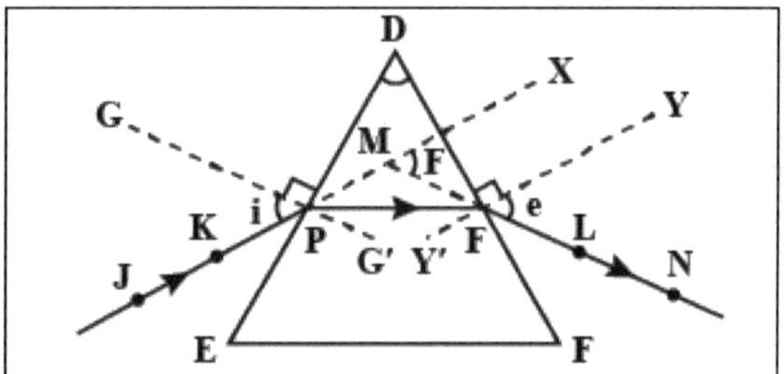

JP - Incident ray ∠i - Angle of incidence
PO - Refracted ray ∠r - Angle of refraction
ON - Emergent ray ∠e - Angle of emergence
D - Angle of the prism ∠F - Angle of deviation

In the figure, D is the angle of the prism.

Snell's law mentions that, when traveling from a rarer medium to a denser medium light bends towards the normal, and bends away from the normal when traveling from denser medium to rarer medium. Here glass is denser than air, and therefore, when the light ray falls on the surface of the prism, the ray of light bends towards the normal. Hence according to the figure, when ray JP falls on the surface of the prism ,it bends towards the normal GP.

The emergent ray ON while moving from glass to air, it bends away from the normal. The angle of deviation, ∠XFN gives us the degree at which the emergent ray has deviated from the incident ray. The angle of deviation is minimum when the angle of incidence is equal to angle of emergence. In the diagram given, ∠JPG = ∠YPN and ∠XFN is the angle of deviation, which is minimum. In this case the refracted ray PO is parallel to side EF.

This is how when passing through a prism, dispersion of light (white light) happens wherein it splits into it's seven constituent colours. Different colors of different wavelength experience different degree of deviation and hence when put through refraction, white light splits into its constituent colors.

Dispersion of Light and Rainbow

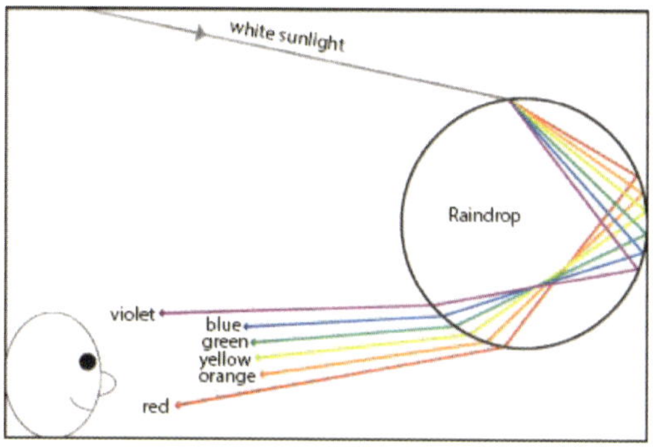

A rainbow is formed when it rains and there's sunshine simultaneously. Each raindrop is made of a different shape and has a different consistency as compared to a glass prism, still it affects light in a similar fashion. One can see the seven constituent colors (red, orange, yellow, green, blue, indigo and violet) when white light (sunlight) hits a collection of raindrops at a fairly low angle.

When the white sunlight travels from air into the drop of water, the constituent colors of light slow down to varying speed and frequencies. When violet light enters the raindrop it bends at a relatively sharp angle. Some of the light then passes back out into the air, and the rest is reflected back. Some of that reflected light passes out of the drop, bending as it moves back into the air again.

This way, each raindrop disperses the white sunlight into its constituent colors. We see wide bands of color, as if different rainy areas were dispersing a different single color since we only see one color from each raindrop.

Sometimes we also see a double rainbow, a sharp rainbow with a fainter rainbow above it. The fainter rainbow occurs in the same manner as the sharper rainbow, but here the light is reflected twice inside the raindrop instead of reflecting once. Due to this double reflection, the light exits the raindrop at a different angle, so we see it above the sharper rainbow. And if you observe carefully, you'll notice that the colors in the secondary rainbow are in the reverse order of the primary rainbow.

Total Internal Reflection

A good-quality mirror may reflect more than 90% of the light that falls on it, absorbing the rest. But it would be useful to have a mirror that reflects all of the light that falls on it. Interestingly, we can produce total reflection using an aspect of refraction.

Consider what happens when a ray of light strikes the surface between two materials, such as is shown in figure (a) below. Part of the light crosses the boundary and is refracted; the rest is reflected. If, as shown in the figure, the index of refraction for the second medium is less than for the first, the ray bends away from the perpendicular. (Since $n_1 > n_2$, the angle of refraction is greater than the angle of incidence—that is, $\theta_1 > \theta_2$.) Now imagine what happens as the incident angle is increased. This causes θ_2 to increase also. The largest the angle of refraction θ_2 can be is 90°, as shown in figure. The critical angle θ_c for a combination of materials is defined to be the incident angle θ_1 that produces an angle of refraction of 90°. That is, θ_c is the incident angle for which $\theta_2 = 90°$. If the incident angle θ_1 is greater than the critical angle, as shown in figure, then all of the light is reflected back into medium 1, a condition called total internal reflection.

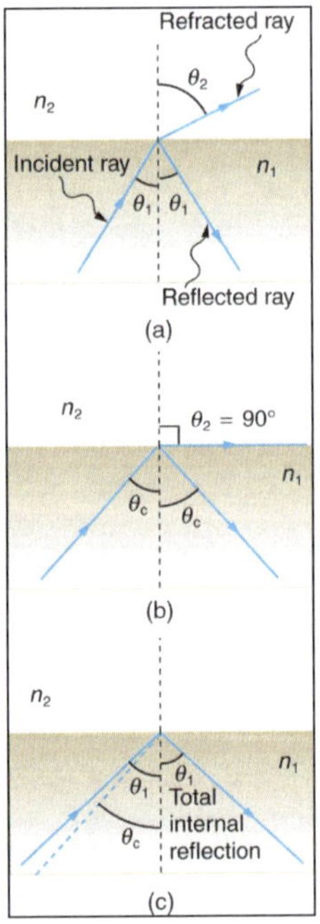

(a) A ray of light crosses a boundary where the speed of light increases and the index of refraction decreases. That is, $n_2 < n_1$. The ray bends away from the perpendicular. (b) The critical angle θ_c is the one for which the angle of refraction is. (c) Total internal reflection occurs when the incident angle is greater than the critical angle.

Snell's law states the relationship between angles and indices of refraction. It is given by,

$$n_1 \sin\theta_1 = n_2 \sin\theta_2.$$

When the incident angle equals the critical angle $(\theta_1 = \theta_c)$, the angle of refraction is $90° (\theta_2 = 90°)$. Noting that $\sin 90° = 1$, Snell's law in this case becomes,

$$n_1 \sin\theta_1 = n_2.$$

The critical angle θ_c for a given combination of materials is thus,

$$\theta_c = \sin^{-1}\left(\frac{n_2}{n_1}\right)$$

for $n_1 > n_2$.

Total internal reflection occurs for any incident angle greater than the critical angle θ_c, and it can only occur when the second medium has an index of refraction less than the first. Note the above equation is written for a light ray that travels in medium 1 and reflects from medium 2, as shown in the figure.

Fiber Optics: Endoscopes to Telephones

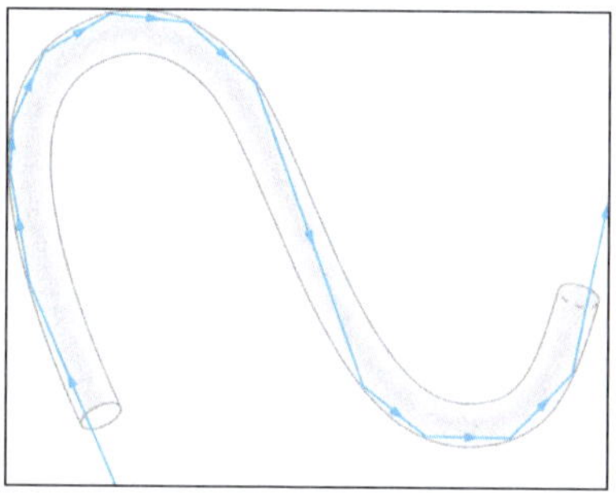

Figure: Light entering a thin fiber may strike the inside surface at large or grazing angles and is completely reflected if these angles exceed the critical angle. Such rays continue down the fiber, even following it around corners, since the angles of reflection and incidence remain large.

Fiber optics is one application of total internal reflection that is in wide use. In communications, it is used to transmit telephone, internet, and cable TV signals. *Fiber optics* employs the transmission of light down fibers of plastic or glass. Because the fibers are thin, light entering one is likely to strike the inside surface at an angle greater than the critical angle and, thus, be totally reflected. The index of refraction outside the fiber must be smaller than inside, a condition that is easily satisfied by coating the outside of the fiber with a material having an appropriate refractive index. In fact, most fibers have a varying refractive index to allow more light to be guided along the fiber through total internal refraction. Rays are reflected around corners as shown, making the fibers into tiny light pipes.

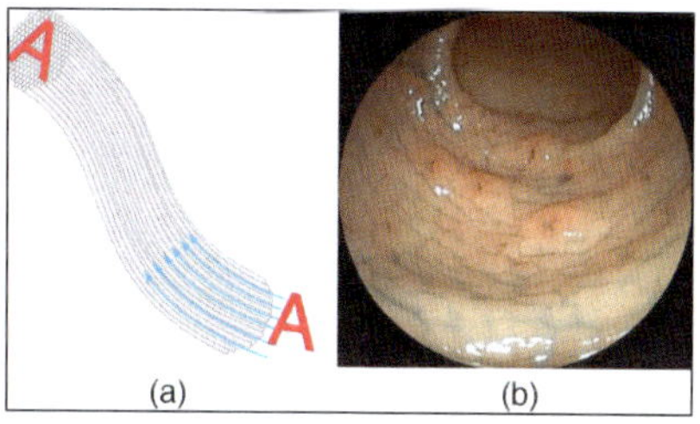

Figure: (a) An image is transmitted by a bundle of fibers that have fixed neighbors. (b) An endoscope is used to probe the body, both transmitting light to the interior and returning an image such as the one shown.

Bundles of fibers can be used to transmit an image without a lens, as illustrated in figure. The output of a device called an *endoscope* is shown in figure. Endoscopes are used to explore the body through various orifices or minor incisions. Light is transmitted down one fiber bundle to illuminate internal parts, and the reflected light is transmitted back out through another to be observed. Surgery can be performed, such as arthroscopic surgery on the knee joint, employing cutting tools attached to and observed with the endoscope. Samples can also be obtained, such as by lassoing an intestinal polyp for external examination.

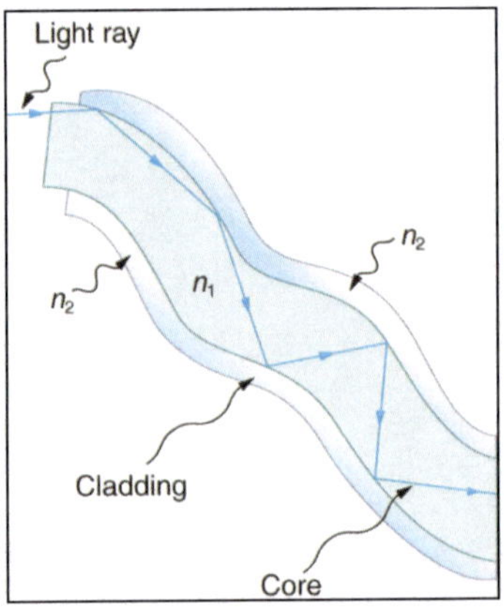

Figure: Fibers in bundles are clad by a material that has a lower index of refraction than the core to ensure total internal reflection, even when fibers are in contact with one another. This shows a single fiber with its cladding.

Fiber optics has revolutionized surgical techniques and observations within the body. There are a host of medical diagnostic and therapeutic uses. The flexibility of the fiber optic bundle allows it to navigate around difficult and small regions in the body, such as the intestines, the heart, blood vessels, and joints. Transmission of an intense laser beam to burn away obstructing plaques in major arteries as well as delivering light to activate chemotherapy drugs are becoming commonplace. Optical fibers have in fact enabled microsurgery and remote surgery where the incisions are small and the surgeon's fingers do not need to touch the diseased tissue.

Fibers in bundles are surrounded by a cladding material that has a lower index of refraction than the core. The cladding prevents light from being transmitted between fibers in a bundle. Without cladding, light could pass between fibers in contact, since their indices of refraction are identical. Since no light gets into the cladding (there is total internal reflection back into the core), none can be transmitted between clad fibers that are in contact with one another. The cladding prevents light from escaping out of the fiber; instead most of the light is propagated along the length of the fiber, minimizing the loss of signal and ensuring that a quality image is formed at the other end. The cladding and an additional protective layer make optical fibers flexible and durable.

Special tiny lenses that can be attached to the ends of bundles of fibers are being designed and fabricated. Light emerging from a fiber bundle can be focused and a tiny spot can be imaged. In some cases the spot can be scanned, allowing quality imaging of a region inside the body. Special minute optical filters inserted at the end of the fiber bundle have the capacity to image tens of microns below the surface without cutting the surface—non-intrusive diagnostics. This is particularly useful for determining the extent of cancers in the stomach and bowel.

Most telephone conversations and Internet communications are now carried by laser signals along optical fibers. Extensive optical fiber cables have been placed on the ocean floor and underground to enable optical communications. Optical fiber communication systems offer several advantages over electrical (copper) based systems, particularly for long distances. The fibers can be made so transparent that light can travel many kilometers before it becomes dim enough to require amplification—much superior to copper conductors. This property of optical fibers is called low loss. Lasers emit light with characteristics that allow far more conversations in one fiber than are possible with electric signals on a single conductor. This property of optical fibers is called high bandwidth. Optical signals in one fiber do not produce undesirable effects in other adjacent fibers. This property of optical fibers is called reduced crosstalk.

Corner Reflectors and Diamonds

A light ray that strikes an object consisting of two mutually perpendicular reflecting surfaces is reflected back exactly parallel to the direction from which it came. This is true whenever the reflecting surfaces are perpendicular, and it is independent of the angle of incidence. Such an object, shown in Figure, is called a corner reflector, since the light bounces from its inside corner. Many inexpensive reflector buttons on bicycles, cars, and warning signs have corner reflectors designed to return light in the direction from which it originated. It was more expensive for astronauts to place one on the moon. Laser signals can be bounced from that corner reflector to measure the gradually increasing distance to the moon with great precision.

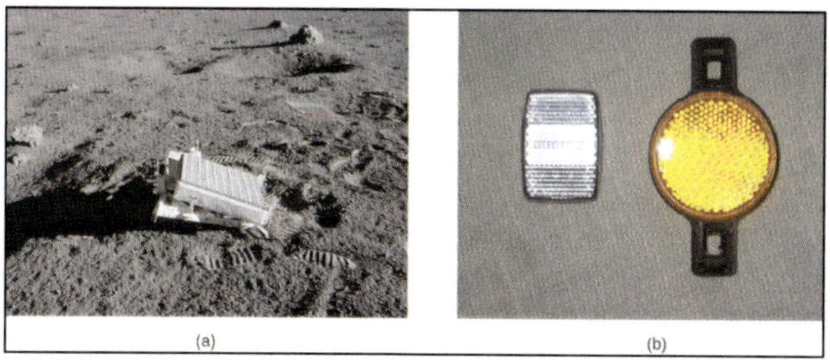

Figure: (a) Astronauts placed a corner reflector on the moon to measure its gradually increasing orbital distance. (b) The bright spots on these bicycle safety reflectors are reflections of the flash of the camera that took this picture on a dark night.

Corner reflectors are perfectly efficient when the conditions for total internal reflection are satisfied. With common materials, it is easy to obtain a critical angle that is less than 45°. One use of these perfect mirrors is in binoculars, as shown in figure. Another use is in periscopes found in submarines.

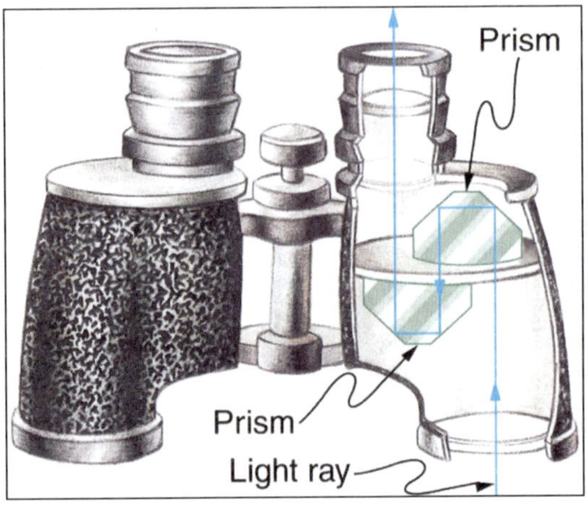

Figure: These binoculars employ corner reflectors with total internal reflection to get light to the observer's eyes.

The Sparkle of Diamonds

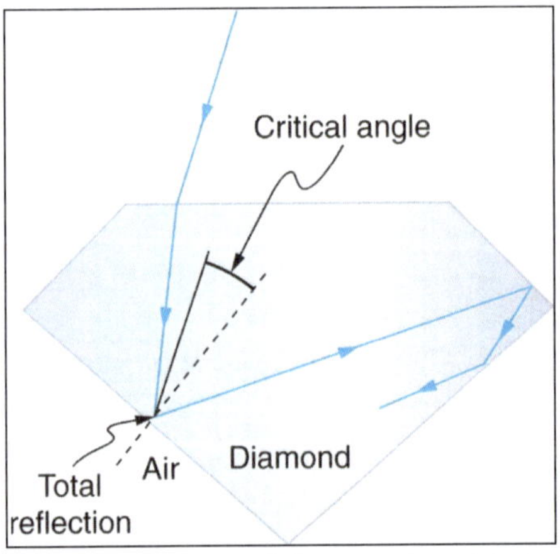

Figure: Light cannot easily escape a diamond, because its critical angle with air is so small. Most reflections are total, and the facets are placed so that light can exit only in particular ways—thus concentrating the light and making the diamond sparkle.

Total internal reflection, coupled with a large index of refraction, explains why diamonds sparkle more than other materials. The critical angle for a diamond-to-air surface is only 24.4°, and so when light enters a diamond, it has trouble getting back out. Although light freely enters the diamond, it can exit only if it makes an angle less than 24.4°. Facets on diamonds are specifically intended to make this unlikely, so that the light can exit only in certain places. Good diamonds are very clear, so that the light makes many internal reflections and is concentrated at the few places it can exit—hence the sparkle. (Zircon is a natural gemstone that has an exceptionally large index of refraction, but not as large as diamond, so it is not as highly prized. Cubic zirconia is manufactured and has an even higher index of refraction (≈ 2.17), but still less than that of diamond.) The colors you see emerging from a sparkling diamond are not due to the diamond's color, which is usually nearly colorless.

Scattering of Light

The scattering of light is an important part of our daily life, although we didn't realize its importance. Scattering of light is different from reflection, as in reflection the radiation is defected in one direction while in scattering every object or particle can scatter light and illuminates them in all direction. When a parallel beam of light passes through any particle present in air or gas; the particles present in air scatters the light beam in all directions besides its incident direction. This phenomenon of, light striking the particles present in the air and after absorbing some light it radiates it in different directions except its incident direction, is called as "Scattering of light". The loss of energy in a light beam after scattering can be calculated by the strength of scattering, its value depends on the wavelength of the light and the size of the particle which causes scattering. This process of scattering of light can also be shown with the help of an example of light rays falling from the sun, as when the rays of the sun enter the earth's atmosphere there are various small particles present in the air. These particles collide with the sun rays falling and while absorbing some light making them scatter/ defect in different directions besides its incident direction. Basically, we can say that a ray of light is defected from its straight path due to some irregularities present in the medium, particles or due to the interference between the two media.

There are several ways how the scattering of light takes place, but two main examples are:

- Random refection from a rough surface.
- Refection (refraction or diffraction) through impurities present in the volume.

Random Refection from a Rough Surface

All surfaces are rough; the roughness of the surface is related to the wavelength of the light ray. As the surface having more roughness absorbs more energy from the light ray and will scatter it in different directions depending on the wavelength of the light ray. The rough surface of a car or pieces of jewellery can be taken as a good example of scattering through a rough surface.

Refection through Impurities Present in Volume

This type of scattering through impurities present in the volume helps in the medical types of equipment. Thompson mechanism, which is also known as elastic scattering, is a good example of this type of scattering which is used in medical X-ray photographs. This type of scattering occurs where light is scattered by charged particles by levelling the energy of the incident radiation and the rest energy of the charged particle present. The passing of light ray through any liquid in which light beam scatters only with the tiny fat droplets is a good example of this type of scattering.

Single and Multiple Scattering

When the scattering of light takes place with only a single localized scattering center then the phenomenon is known as single scattering, it is generally treated as a random phenomenon as it has only a single scattering center which can do a single scattering event at a particular time. While those scattering which takes place with many localized scattering centers, a large number of localized center involves a large number of combined results which give more number of scattered light to the observer. The single type of scattering is not always random, as they can be intentional sometime in the case of a laser beam, which can be well controlled for scattering to a single point for an instance, along with the radar scattering where the targets tend to be macroscopic in size.

Types of Scattering of Light

- Rayleigh or selective scattering.
- Mie scattering.
- Electromagnetic scattering.

Rayleigh or Selective Scattering

Rayleigh scattering of light is a type of elastic scattering as the particle from which the scattering has to be done depends on the wavelength of light. Depending on the wavelength, certain particles are more effective than others which scatter light having more wavelength, as the particles like molecules of oxygen, nitrogen having small size scatters light with a shorter wavelength (blue or violet) in different directions. The blue sky on a clear sunny day is also the result of Rayleigh scattering by the air molecules. The blue light having shorter wavelength appears to scatter from the upper atmosphere for about 10 times which is much larger than the red light having a larger wavelength. Hence, the blue light having shorter wavelength collides with the air molecules and is scattered to the eyes seeing the sky, making it blue. On the other hand, the red light with a higher wavelength goes largely unscattered to the sky. Scattering of optical signals through optical fibres are also included in this type of scattering.

Mie Scattering

Mie scattering is also a type of elastic scattering mechanism, in which the size of the molecules is greater than the wavelength of the light which results in the non-uniform scattering of light. This type of scattering is not much dependent on wavelength as the size of the molecule which scatters light is more important. Due to this process, the clouds having water droplets looks white. The scattering efficiency of the small molecules becomes less in the atmosphere with the wavelength of the white light. This shows that the light ray which enters in the clouds get scattered by water droplets for all wavelength and no light of visible wavelength is left in the cloud making the cloud look white. When the clouds become full of water droplets no light is taken in the cloud for scattering making it looks darker. The light rays falling on the earth's surface results in the different types of scattering, making the color of the sky blue and making the color of the clouds white. The white color of fog and clouds both are the results of "Mie scattering".

Electromagnetic Scattering

It is the most common form of scattering as it includes electromagnetic waves. It generally includes two types of scattering which are elastic light and inelastic light scattering. Elastic light scattering includes Rayleigh scattering or Mie scattering while inelastic scattering includes Raman scattering, inelastic x-ray scattering, Compton scattering, and Brillouin scattering.

The intensity of the light scattered from the molecules generally depends on two factors which are the wavelength of the light to be scattered and the size of the molecules due to which the light falling on the earth's surface is scattered. The light ray falling on the earth's surface having a shorter wavelength and high frequency tend to scatter more because of the intersection with the particles and the waviness of the line. While the light rays falling on the earth's surface having low frequency and longer wavelength tend to scatter less because they move in a straighter path which makes the possibility of colliding with the particles less.

The probability of scattering and λ the wavelength of the light can give the relation as follows:

$$P \propto 1/\lambda 4$$

Where P is the probability of scattering.

And, λ is the wavelength of the light.

The above relation between the probability of scattering and the wavelength of the light shows that the probability of scattering is inversely proportional to the fourth power of the wavelength and the probability of scattering will be higher for the shorter wavelength of the light.

Transmission, Absorption, Emission and Reflection

Transmission is the term used to describe the process by which incident radiant flux leaves a surface or medium from a side other than the incident side, usually the opposite side. The spectral transmittance $\tau(\lambda)$ of a medium is the ratio of the transmitted spectral flux $\Phi_{\lambda t}$ to the incident spectral flux $\Phi_{\lambda i}$, or

$$\tau(\lambda) = \frac{\Phi_{\lambda t}}{\Phi_{\lambda i}}$$

The transmittance τ is the ratio of the transmitted flux Φ_t to the incident flux Φ_i or

$$\tau = \frac{\int_0^\infty \tau(\lambda)\Phi_{\lambda i} d\lambda}{\int_0^\infty \Phi_{\lambda i} d\lambda} \neq \int_\lambda \tau(\lambda) d\lambda$$

Note that the integrated transmittance is not the integral over wavelength of the spectral transmittance, but must be weighted by a source function Φ_λ as shown.

The transmittance may also be described in terms of radiance as follows

$$\tau = \frac{\int_0^\infty \int_\Omega L_{\lambda i} d\Omega_t d\lambda}{\int_0^\infty \int_\Omega L_{\lambda i} d\Omega_i d\lambda}$$

Where $L_{\lambda i}$ represents the spectral radiance $L_{\lambda i}(\lambda;\theta_i,\phi_i)$ incident from direction (θ_i,ϕ_i), $L_{\lambda i}$ represents the spectral radiance $L_{\lambda i}(\lambda;\theta_t,\phi_t)$ transmitted in direction (θ_t,ϕ_t), and $d\Omega$ is the elemental projected solid angle $\sin\theta\cos\theta\, d\theta\, d\phi$.

The bidirectional transmittance distribution function (BTDF , symbol f_t) relates the transmitted radiance to the radiant incidence as,

$$f_t(\lambda;\theta_i,\phi_i) \equiv \frac{dL_{\lambda t}}{dL_{\lambda i} d\Omega_i} = \frac{dL_{\lambda t}}{dE_{\lambda i}}\left(sr^{-1}\right)$$

Geometrically, transmittance can be classified as specular, diffuse, or total, depending upon whether the specular (regular) direction, all directions other than the specular, or all directions are considered.

Absorptance

Absorption is the process by which incident radiant flux is converted to another form of energy, usually heat. Absorptance is the fraction of incident flux that is absorbed. The absorptance a of an element is defined by $\alpha = \Phi_a / \Phi_i$. Similarly , the spectral absorptance $\alpha(\lambda)$ is the ratio of spectral power absorbed $\Phi_{\lambda a}$ to the incident spectral power $\Phi_{\lambda i}$,

$$\alpha = \frac{\int_0^\infty \alpha(\lambda)\Phi_{\lambda i} d\lambda}{\int_0^\infty \Phi_{\lambda i} d\lambda} \neq \int_\lambda \alpha(\lambda) d\lambda$$

An absorption coefficient α'(cm^{-1} or km^{-1}) is often used in the expression $\tau_i = e^{-\alpha' t}$, where τ_i is internal transmittance and t is pathlength (cm or km).

Reflectance

Reflection is the process where a fraction of the radiant flux incident on a surface is returned into the same hemisphere whose base is the surface and which contains the incident radiation. The reflection can be specular (in the mirror direction), diffuse (scattered into the entire hemisphere), or a combination of both.

The most general definition for reflectance ρ is the ratio of the radiant flux reflected Φ_r to the incident radiant flux Φ_i, or

$$\rho = \frac{\Phi_r}{\Phi_i}$$

Spectral reflectance is similarly defined at a specified wavelength λ as

$$\rho(\lambda) = \frac{\Phi_{\lambda r}}{\Phi_{\lambda i}}$$

(Spectral) reflectance factor (symbol R) is the ratio of (spectral) flux reflected from a sample to the (spectral) flux which would be reflected by a perfect diffuse (lambertian) reflector.

No single descriptor of reflectance will suffice for the wide range of possible geometries.

The fundamental geometric descriptor of reflectance is the bidirectional reflectance distribution function (BRDF, symbol f_r). It is defined as the differential element of reflected radiance dL_r in a specified direction per unit differential element of radiant incidence dE_i, also in a specified direction, and carries unit of sr^{-1}.

$$f_r(\theta_i, \phi_i, \theta_r, \phi_r) = \frac{dL_r(\theta_i, \phi_i, \theta_r, \phi_r; E_i)}{dE_i(\theta_i, \phi_i)} \left[sr^{-1} \right]$$

The polar angle θ is measured from the surface normal and the azimuth angle f is measured from an arbitrary reference in the surface plane, most often the plane containing the incident beam. The subscripts i and r refer to the incident and reflected beams, respectively.

By integrating over varying solid angles, Nicodemus et al., based upon earlier work by Judd, defined nine goniometric reflectances, and by extension, nine goniometric reflectance factors.

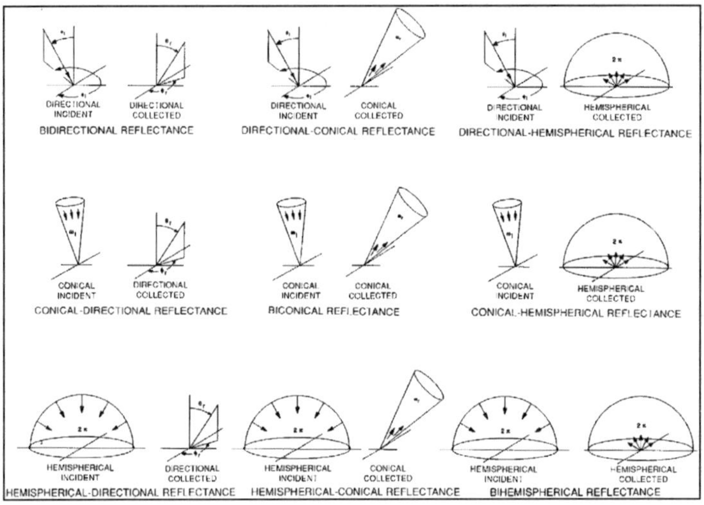

Figure: Nine geometrical definitions of reflectance

Emittance

Emittance (ε) is the ratio of the radiance of an object or surface to the radiance of a blackbody (planckian radiator) at the same temperature. It is therefore dimensionless and can assume values between 0 and 1 for thermal radiators at equilibrium. Spectral emittance $\varepsilon(\lambda)$ is the emittance at a given wavelength. If a radiator is neutral with respect to wavelength, with a constant spectral emittance less than unity, it is called a graybody.

$$\varepsilon \frac{L}{L^{bb}} \quad \varepsilon(\lambda) = \frac{L_\lambda}{L_\lambda^{bb}}$$

Directional emittance $\varepsilon(\theta,\phi)$ is defined by

$$\varepsilon(\theta,\phi) = \frac{L(\theta,\phi)}{L^{bb}}$$

Note that if the body is nongray, its emittance is dependent upon temperature inasmuch as the integral must be weighted by the source (Planck) function.

$$\varepsilon = \frac{\int_0^\infty \varepsilon(\lambda) L_\lambda^{bb} d\lambda}{\int_0^\infty L_\lambda^{bb} d\lambda} = \frac{1}{\pi} \cdot \frac{\int_0^\infty \varepsilon(\lambda) L_\lambda^{bb} d\lambda}{\sigma T^4}$$

Kirchhoff's Law

In a closed system at thermal equilibrium, conservation of energy necessitates that emitted and absorbed fluxes are equal. Since the radiation field in such a system is isotropic (the same in all directions), the directional spectral emittance and the directional spectral absorptance must be equal, i. e.,

$$\varepsilon(\lambda;\theta,\phi)=\alpha(\lambda;\theta,\phi)$$

This statement was first made by Kirchhoff. Strictly, this equation holds for each orthogonal polarization component, and for it to be valid as written, the total radiation must have equal orthogonal polarization components. Kirchhoff's law is often simplified to the declaration $\alpha=\varepsilon$; however, this is not a universal truth; it may only be applied under a limited set of conditions. The geometrical and spectral averaging (integration) is governed by a specific set of rules as demonstrated by Siegel and Howell.

Relationship between Transmittance, Reflectance and Absorptance

Radiant flux incident upon a surface or medium undergoes transmission, reflection, and absorption. Application of conservation of energy leads to the statement that the sum of the transmission, reflection, and absorption of the incident flux is equal to unity, or

$$\alpha+\tau+\rho=1$$

In the absence of nonlinear effects (i. e., Raman effect, etc.),

$$\alpha(\lambda)+\tau(\lambda)+\rho(\lambda)=1$$

If the situation is such that one of the above Kirchhoff-type relations is applicable, then emittance ε may be substituted for absorptance α in the previous equations, or

$$\varepsilon=1-\tau-\rho \qquad \varepsilon(\lambda)=1-\tau(\lambda)-\rho(\lambda)$$

Measurement of Transmittance

A knowledge of the transmission of optical materials and elements, gaseous atmospheres, and various liquids is necessary throughout the realm of optics. Most of these measurements are made with commercial spectrophotometers. It is beyond the scope to discuss thc design and operation of spectrophotometric equipment beyond sample-handling practices.

Conventional spectrophotometers are of the double-beam configuration, where the output is the ratio of the signal in the sample beam to the signal in the reference beam plotted as a function of wavelength. It is incumbent upon the experimenter to ensure that the only difference between the two beams is the unknown. Therefore, if liquid or gas cells are employed, one should be placed in each beam. For gas cells, an equal amount of carrier gas should be injected into each cell, with the unknown to be sample placed in only one cell, destined for the sample beam. For liquids, an equal amount of solute should be placed in each cell. A critical issue with liquid and solid samples is the beam geometry. Most spectrophotometers feature converging beams in the sample space. If the optical path (the product of index of refraction and actual distance) is not identical for each beam, a systematic difference is presented to either the entrance slit or the detector. In addition, some specimens (e. g., interference filters) are susceptible to errors when measured in a converging beam. Most instruments also have a single monochromator which is susceptible to stray radiation, the limiting factor when trying to make measurements of samples that are highly absorbing in one spectral region and transmitting in another. Some recent instruments feature linear detector arrays along with single monochromators to allow the acquisition of the entire spectrum in several milliseconds; these are particularly applicable to reaction rate studies.

Conventional double-beam instruments are limited by these factors to uncertainties on the order of 0.1 percent. For lower uncertainties, the performance deficiencies found in double-beam instruments can largely be overcome by the use of a single-beam architecture. The mode of operation is sample-in – sample-out. If the source is sufficiently stable with time, the desired spectral range can be scanned without the sample, then rescanned with the sample in place. Otherwise, the spectrometer can be set at a fixed wavelength and alternate readings with and without the sample in place must be made. Care should be taken to ensure that the beam geometry is not altered between sequential readings.

To achieve the ultimate in performance from conventional spectrophotometry, several design characteristics should be included. A double monochromator is essential to minimize stray light. The beam geometry in the sample compartment should be highly collimated to avoid focus shifts with optically thick samples. Some form of beam integration, such as an integrating sphere or other diffuser, should be employed to negate the effects of nonuniform detectors and beam shifts. An exemplary instrument is the high-accuracy spectrophotometer developed by the National Institute for Standards and Technology (NIST), described in Mielenz et al., Mielenz et al., Venable et al., Eckerle, and Eckerle et al. A particularly useful review is Eckerle et al. Similar laboratory instruments have also been built elsewhere by Clarke, Freeman, and Zwinkles and Gignac.

Numerous other instruments have been described in the literature; some have been designed for singular or limited purposes while others have a more universal appeal. Use of integrating spheres is common, both for the averaging effects and for the isolation of the specular and diffuse components, as shown in figure Several useful instruments are described by Karras, Taylor, Zerlaut and Anderson, Clarke and Larkin, and Kessel.

Conventional instruments lack a wide dynamic range because there are simply not enough photons available in a narrow bandpass in a reasonable time. Solutions include Fourier transform spectrometers with a large multiplex advantage, the use of tunable lasers, and heterodyne spectrometry.

Simple instruments can be purchased or constructed for specific purposes. For example, solar transmittance can be determined using either the natural sun (if available) or simulated solar radiation as the source. A limited degree of spectral isolation can be achieved with an abridged spectrophotometer using narrow bandpass interference or glass absorption filters.

Several publications have suggested methods for making accurate and repeatable measurements including Hughes, Mielenz, Venable and Hsia, Burke and Mavrodineanu, ASTM F768, ASTM E971, ASTM E903, and ASTM E179. Calibration and performance assessment of spectrophotometers includes photometric accuracy, linearity, stray light analysis, and wavelength calibration. Particular attention should be paid to luminescent samples that absorb radiant energy in one spectral region and re-emit it at longer wavelengths.

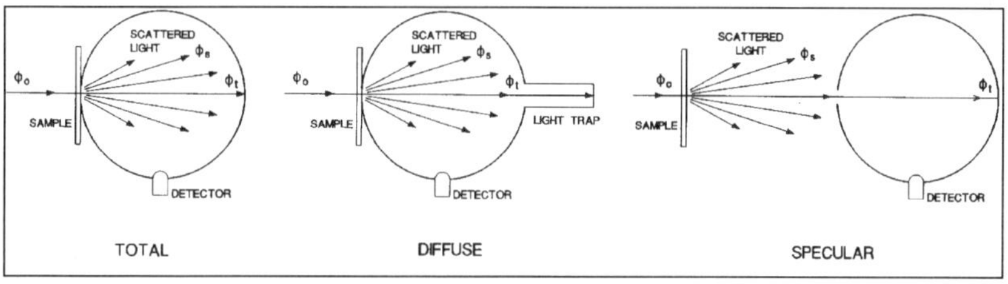

Figure: Measurement of total, diffuse, and specular transmittance using an integrating sphere.

Standards of spectral transmittance are available as Standard Reference Materials from NIST. These take the form of metal-on-glass, metal-on-quartz, and solid glass filters. Most are used for verifying the photometric scale or for checking the wavelength calibration of a recording spectrophotometer. Descriptions of their development and use are given in Mavrodineanu and Baldwin, Mavrodineanu and Baldwin, Eckerle et al., and Hsia. The standardization laboratories of several countries occasionally conduct international intercomparisons of traveling standards.

Measurement of Absorptance

In most cases, absorptance is not directly measured, but is inferred from transmission measurements, with appropriate corrections for reflection losses. These corrections can be calculated from the Fresnel equations if the surfaces are polished and the index of refraction is known. For materials where the absorption is extremely small, this method is unsatisfactory, as the uncertainties are dominated by the reflection contribution.

Measurement of Reflectance

Instrumentation for the measurement of reflectance takes many forms. Only a few of the definitions for reflectance and reflectance factor have been adopted as standard configurations. The biconical configuration with small solid angles is most suited to a measurement of specular (regular, in the mirror direction) reflection. A simple reflectometer for the absolute measurement of specular reflectance was devised by Strong and is shown schematically in figure below. Numerous detail improvements have been made on this fundamental design, including the use of averaging spheres. Designs range from simple to complex. Some reflectometers have been built specifically to measure at normal incidence.

The characterization of appearance of materials involves measurements of reflectance, both diffuse and specular. Numerous procedures and instruments have been devised for goniophotometry, the measurement of specular gloss with biconical geometry. Measurements are made at several angles from normal (20°, 30°, 45°, 60°, 75°, and 85°) depending upon the material under scrutiny.

In the visible and near-IR spectral regions, the integrating sphere is the instrument of choice for both specular and diffuse specimens. Many papers have been written detailing instruments, methods, and procedures, some of which are shown in Figure The specular component of the reflected flux can be included to determine the total reflectance or excluded to measure just the diffuse component. The angle of incidence can be varied by placing the sample at the center of the sphere. Some of these methods have been incorporated into standard methods and practices, such as ASTM C523, ASTM E429, ASTM E903, CIE, and IES. Most integrating sphere measurements require reference to some form of an artifact standard, but the double- sphere method produces absolute diffuse reflectance. Lindberg has demonstrated a method to scale relative measurements to absolute.

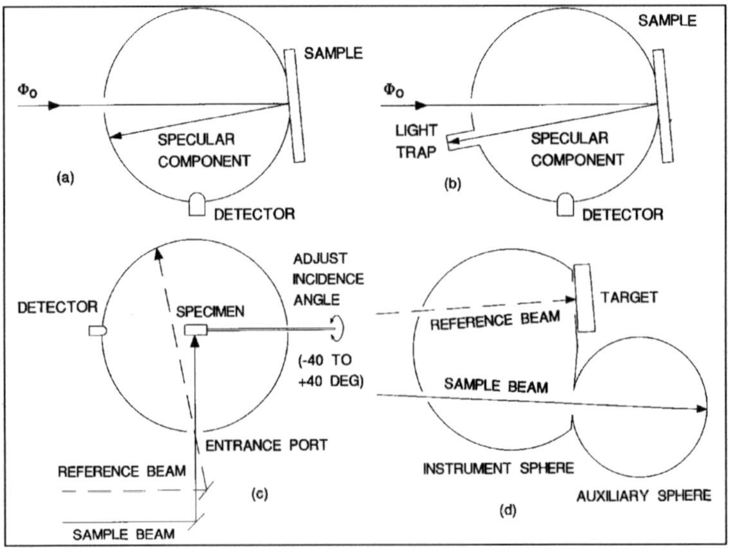

Figure: Measurement of diffuse reflectance using an integrating sphere.

Alternative forms of hemispherical irradiation and collection have been described, several of which are shown in figure. Specular hemispherical, paraboloidal, and ellipsoidal collectors have been used, particularly in those spectral regions where integrating sphere coatings are difficult to obtain. The Helmholtz reciprocity principle has been invoked to demonstrate the reversibility of the source and the collector. Hemispherical irradiation has also been employed by placing a cooled sample coplanar with the wall of a furnace and comparing the radiance of the sample with that of the wall of the furnace as shown in figure below.

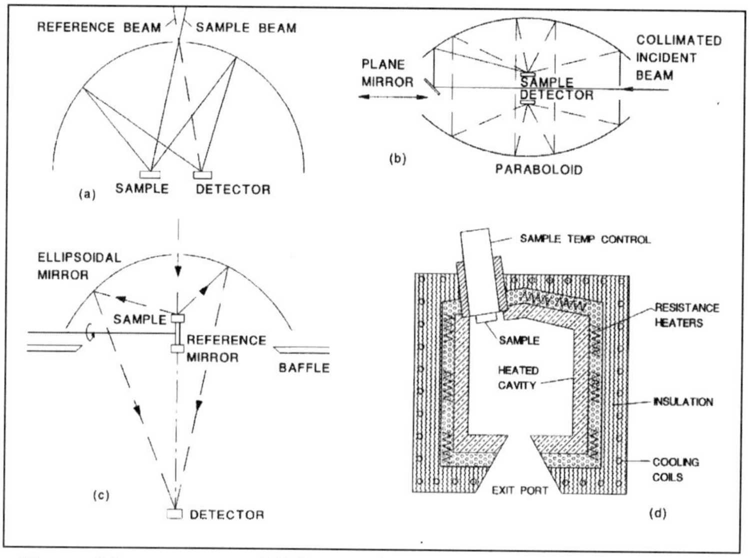

Figure: Measurement of diffuse reflectance using alternate methods.

The procedures and instrumentation for the measurement of reflectance factor are identical with those for diffuse reflectance using the 0°/45° or 45°/0° geometry with annular, circumferential, or uniplanar illumination or viewing. Reference is made to a white reflectance standard characterized for reflectance factor, which must be compared with a perfect diffuse reflector.

Orbiting sensors measure the radiance of the earth-atmosphere system with some known geometry (generally nadir) and in well-defined wavelength bands. The quantity of interest is reflectance, as it is related to factors such as crop assessment, mineralization, etc. Corrections must be made for the atmospheric absorption, emission, and scattering and for the BRDF (bidirectional reflectance distribution function) of the target. BRDF has been characterized in the field using the sun as the source.

Laboratory measurements of BRDF are made using goniometers where the sample- source angle and sample-receiver angle are independently adjustable. Coherent sources (lasers) are employed for the characterization of smooth specimens where a large source power is necessary for adequate SNR for of f-specular angles and where speckle is not a concern. Incoherent sources (xenon arcs, blackbody simulators, or tungsten-halogen) sources are often employed with spectral filters for more diffuse specimens. Similar measurements and techniques are employed to characterize bidirectional transmittance distribution function (BTDF) and bidirectional scattering distribution function (BSDF).

The measurement of retroreflection poses the situation that the return beam coincides with the incident beam. The usual solution is to employ a beam splitter in the system, allowing the incident beam to pass and the return beam to be reflected. This immediately imposes both a significant loss in flux and the situation where the beam reaching the sample is partially polarized, unless a nonpolarizing beamsplitter is employed. In addition, it is imperative that the reflected component of the incident beam be well-trapped, as the radiometer is looking in the same direction. The special vocabulary for retroflection is given in CIE and ASTM E808. Test methods are given in ASTM E810, ASTM E809, and Venable and Johnson.

Most measurements of reflectance are relative and require artifact standards. Specular reflectance measurements are occasionally made using the absolute technique shown in figure, but are more commonly done using a simple reflectance attachment for a commercial spectrophotometer, requiring a calibrated reference.

Diffuse reflectance standards exist in several forms. Ideally, a perfect ($\rho = 1$, lambertian) diffuser would be used, particularly for measuring reflectance factor. Certain materials approach the ideal over a limited angular and wavelength range. Historically, MgO was used in the visible spectrum. It was replaced first by BaSO4 and ASTM E259. This fine, white powder, when pressed to a density of 1g/ cm3, is close to ideal over a wide spectral range. It is not quite lambertian, showing a falloff of BRDF at angles far removed from the specular direction and exhibiting a slight amount of retroreflection. It may also be slightly luminescent when excited by far-ultraviolet.

In the infrared, two materials have proven useful. Flower of sulfur is suitable over the spectral range 1–15 μm . Gold is highly reflective and very stable. To be useful as a diffuse reflectance standard, it must be placed on top of a lambertian surface. Several substrates for gold have been suggested, including sandpaper and flame-sprayed aluminium.

PTFE is a satisfactory laboratory standard but is not well-suited for field use as it is not particularly rugged and is highly adsorbant and therefore subject to contamination. Several solutions have been proposed for working standards, including Eastman integrating sphere paint (BaSO4), Vitriolite tile, and the Russian MS20 and MS14 opal glasses. These materials are, in general, more rugged, stable, and washable than PTFE.

Discussion on the fabrication, calibration, and properties of various diffuse reflectance standards can be found in ASTM E259, Budde, Budde, Egan and Hilgeman, Fairchild and Daoust, Morren et al., TAPPI, and Weidner. International intercomparisons of laboratory standards of diffuse reflectance have been reported in Budde et al. and Weidner and Hsia.

There are no standards available for retroflection. However, NIST offers a Measurements Assurance Program (MAP) to enable laboratories to make measurements consistent with other national standards.

Measurement of Emittance

Measurements of emittance can be done in several ways. The most direct method involves forming a material into the shape of a cavity in such a way that near-blackbody radiation is emitted. A measurement then compares the radiation from a location within the formed cavity to radiation from a flat, outside surface of the material, presumably at the same temperature. The cavity can take the form of a cylinder, cone, or sphere. Similarly, a small-diameter, deep hole can be drilled into a specimen and radiation from the surface compared to radiation from the hole. Care must be taken that the specimen is isothermal and that the reflected radiation is considered. The definitive measurements of several materials, such as tungsten, were determined in this fashion. The significant advantage in this direct method is that it is relative, depending on neither absolute radiometry or thermometry, but only requiring that the radiometer or spectroradiometer be linear over the dynamic range of the measurement. This linearity is also determinable by relative measurements.

If a variable-temperature blackbody simulator and a suitable thermometer are available, the specimen can be heated to the desired temperature T_s and the blackbody simulator temperature T_{bb} can be adjusted such that its (spectral) radiance matches that of the specimen. Then the (spectral) emittance is calculable using the following equations for spectral emittance $\varepsilon(\lambda)$ and emittance ε (for a graybody only).

$$\varepsilon(\lambda) = \frac{e^{c_2/\lambda T_s} - 1}{e^{c_2/\lambda T_{bb}} - 1} \qquad \varepsilon \frac{T_{bb}^4}{T_s^4}$$

If an absolutely calibrated radiometer and a satisfactory thermometer are available, a direct measurement can be made, as L_b is calculable if the temperature is known. Again, the reflected radiation must be considered.

Simple "inspection meter" techniques have been developed, and instrumentation is commercially available to determine the hemispherical emittance over a limited range of temperatures surrounding ambient. These instruments provide a single number, as they integrate both spatially and spectrally.

Measurements of spectral emittance are most often made using spectral reflectance techniques, invoking Kirchhoff's law along with the assumption that the transmittance is zero. The usual geometry of interest is the directional-hemispherical. This can be achieved by either hemispherical irradiation-directional collection, or, using Helmholtz reciprocity, directional irradiation-hemispherical collection. Any standard reflectometry technique is satisfactory.

A direct method for the measurement of total (integrated over all wavelengths) hemispherical emittance is to use a calorimeter as shown in figure below. A heated specimen is suspended in the center of a large, cold, evacuated chamber. The vacuum minimizes gaseous conduction and convection. If the sample suspension is properly designed, the predominant means of heat transfer is radiation. The chamber must be large to minimize the configuration factor between the chamber and the specimen.

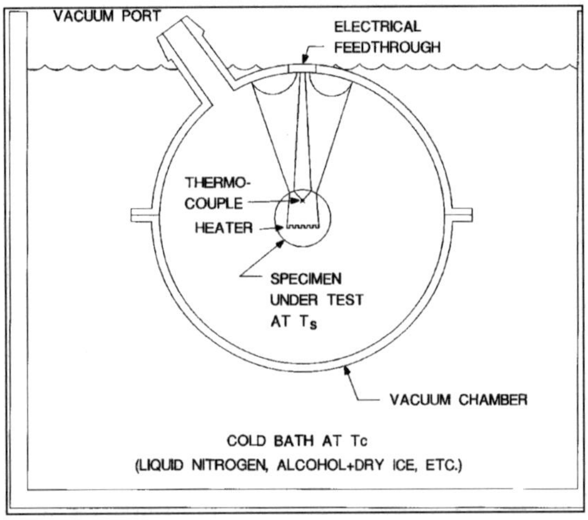

Figure: Calorimetric measurement of total hemispherical emittance.

The chamber is cooled to T_c to reduce radiation from the chamber to the specimen. The equation used to determine emittance ε is,

$$\varepsilon = \frac{P}{\sigma A\left(T_s^4 - T_c^4\right)}$$

where P is the power input to the specimen heater necessary to maintain an equilibrium specimen temperature T_s and A is the specimen area. The equation has been simplified with the aid of the following assumptions: (1) no thermal conduction from the specimen to the chamber, (2) no convective losses, (3) equilibrium has been achieved, and (4) the specimen area is much less than the chamber area. The power can be supplied electrically by means of a known heater or optically via a window in the chamber. In the latter case, a direct measurement of the ratio of solar absorptance α_s to thermal emittance α_T can be directly obtained if the optical source simulates solar radiation. By varying the input power, the emittance can be determined as a function of temperature. There are numerous small corrections to account for geometry, lead conduction, etc.

Several attempts have been made to define and characterize artifact standards of spectral emittance for direct measurements. These were specimens of a thermally stable metal (i. e., Inconel) which were calibrated for emittance as a function of wavelength at several temperatures. No such standards are currently available. Interlaboratory comparisons have been made and reported.

Special problems include measurements at cryogenic temperatures and effects of partially transparent materials.

References

- Images-formed-by-concave-lens, light-reflection-and-refraction: classnotes.org.in, Retrieved 13 June, 2019
- Converging-Lenses-Object-Image-Relations: physicsclassroom.com, Retrieved 11 July, 2019
- Refraction-and-dispersion-of-light, human-eye-and-the-colorful-world, physics: toppr.com, Retrieved 28 March, 2019
- 25-4-total-internal-reflection, physics: courses.lumenlearning.com, Retrieved 13 April, 2019
- Scattering-of-light, physics: vedantu.com, Retrieved 16 May, 2019

5

Polarization and Double Refraction

Polarization is the property of transverse wave that defines the geometrical orientation of the oscillations. Double refraction deals with the anisotropic materials that depend on the polarization and direction of propagation of light. This chapter closely examines all the aspects of polarization and double refraction to provide an extensive understanding of the subject.

Polarization

Polarization is a phenomenon peculiar to transverse waves, i.e., waves that vibrate in a direction perpendicular to their direction of propagation. Light is a transverse electromagnetic wave.

Natural sunlight and almost every other form of artificial illumination transmits light waves whose electric field vectors vibrate in all perpendicular planes with respect to the direction of propagation. When the electric field vectors are restricted to a single plane by filtration, then the light is said to be polarized with respect to the direction of propagation and all waves vibrate in the same plane.

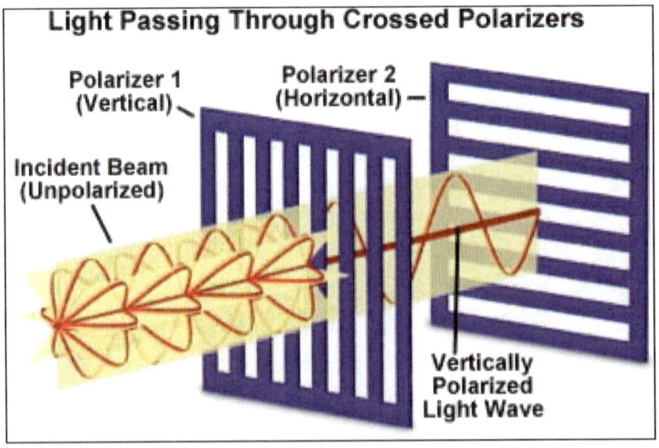

This concept is illustrated in the figure. In this example, the incident light electric field vectors are vibrating perpendicular to the direction of propagation in an equal distribution of all planes before encountering the first polarizer. The polarizers illustrated above are actually filters containing long-chain polymer molecules that are oriented in a single direction. Only the incident light that is vibrating in the same plane as the oriented polymer molecules is absorbed, while light vibrating at right angles to the plane is passed through the first polarizing filter. In Figure, polarizer 1 is oriented vertically to the incident beam so it will pass only the waves that are vertical in the incident beam. The wave passing through polarizer 1 is subsequently blocked by polarizer 2 because the second polarizer is oriented horizontally with respect to the electric field vector in the light wave. The concept of using two polarizers oriented at right angles with respect to each other is commonly termed *crossed polarization* and is fundamental to the practice of polarized light microscopy.

Unpolarized incident light (natural sunlight, for example) is polarized to a certain degree when it is reflected from an insulating surface like water or a highway. In this case, light waves that have the electric field vectors parallel to the surface are reflected to a greater degree than those with different orientations. The optical properties of the insulating surface determine the exact amount of reflected light that is polarized. Mirrors are not good polarizers, although many transparent materials will be very good polarizers, but only if the incident light angle is within certain limits. In this case, the particular angle inducing maximum polarization is known as the *Brewster angle* given by the expression:

$$n = \sin\theta(i)/\sin\theta(r) = \sin\theta(i)/\sin\theta(90\text{-}i) = \tan\theta(i)$$

Where n is the refractive index of the medium, $\theta(i)$ is the angle of incidence, and $\theta(r)$ is the angle of refraction.

This type of polarized light is often termed glare and can be easily demonstrated by viewing the distant part of a highway on a sunny day. Light reflected by the flat surface of a highway is partially polarized with the electric field vectors vibrating in a direction that is parallel to the ground. This light can be blocked by polarizing filters oriented in a vertical direction as illustrated below in figure with a pair of polarized sunglasses.

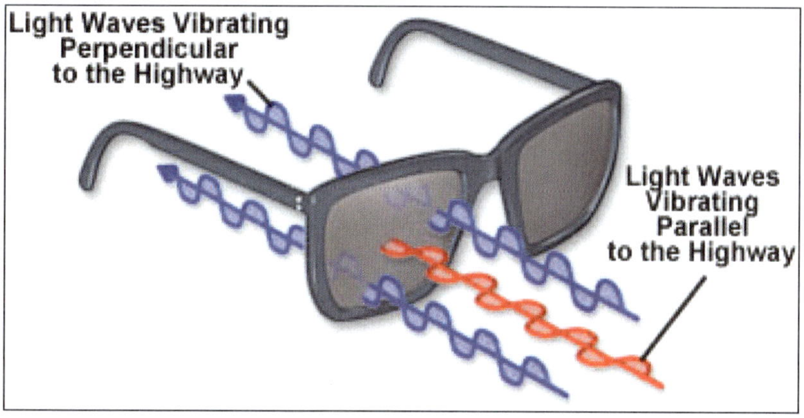

The lenses of the sunglasses have polarizing filters that are oriented vertically with respect to the frames. In the figure above, the blue light waves have their electric field vectors oriented in the same direction as the polarizing lenses and, thus, are passed through. In contrast, the red light wave is perpendicular to the filters and is blocked by the lenses. Polarizing sunglasses are very useful when driving in the sun or at the beach where sunlight is reflected from the surface of the road or water leading to glare that can be almost blinding.

One of the most common uses of polarization today is the liquid crystal display (LCD) used in numerous applications including wrist watches, computer screens, timers, clocks, and many others. These devices are based upon the interaction of rod-like liquid crystalline molecules with an electric field and polarized light waves. The liquid crystalline phase exists in a ground state that is termed cholesteric where the molecules are oriented in layers where each successive layer is slightly twisted to form a spiral pattern. When polarized light waves interact with the liquid crystalline phase the wave is "twisted" by an angle of approximately 90 degrees with respect to the incident wave. This angle is a function of the helical pitch of the cholesteric liquid crystalline phase, which is dependent upon the chemical composition of the molecules (it can be fine-tuned by small changes to the molecules).

An excellent example of the basic application of liquid crystals to display devices can be found in the seven-segment LCD numerical display. Here, the liquid crystalline phase is sandwiched between two glass plates that have electrodes attached similar to those depicted in the illustration below. In figure, the glass plates are drawn with seven black electrodes that can be individually charged (these electrodes are transparent to light in real devices). Light passing through polarizer 1 is polarized in the vertical direction and, when no current is applied to the electrodes, the liquid crystalline phase induces a 90 degree "twist" of the light and it can pass through polarizer 2, which is polarized horizontally and is perpendicular to polarizer 1. This light can then form one of the seven segments on the display.

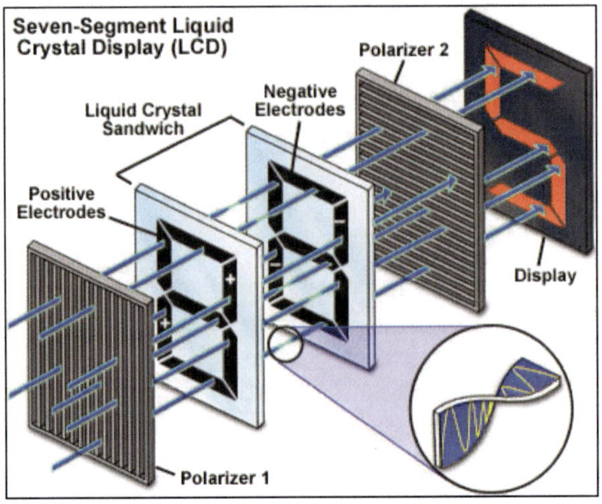

When current is applied to the electrodes, the liquid crystalline phase aligns with the current and loses the cholesteric spiral pattern. Light passing through a charged electrode is not twisted and is blocked by polarizer 2. By coordinating the voltage on the seven positive and negative electrodes, the display is capable of rendering the numbers 0 through 9. In this example the upper right and lower left electrodes are charged and block light passing through them, allowing formation of the number "2".

Polarization of light is very useful in many aspects of optical microscopy. The microscope configuration uses crossed polarizers where the first polarizer (termed: the polarizer) is placed below the sample in the light path and the second polarizer (termed: the analyzer) is placed above the sample, between the objective and the eyepieces. With no sample on the microscope stage, the light polarized by the polarizer is blocked by the analyzer and no light is visible. When samples that are birefringent are viewed on the stage between crossed polarizers, the microscopist can visualize aspects of the samples through light rotated by the sample and then able to pass through the analyzer.

Production of Polarized Light

The various methods of producing polarized light are:

Polarization by use of a Polaroid Filter

The most common method of polarization involves the use of a Polaroid filter. Polaroid filters are made of a special material that is capable of blocking one of the two planes of vibration of an electromagnetic wave. (Remember, the notion of two planes or directions of vibration is merely a simplification that helps us to visualize the wavelike nature of the electromagnetic wave.) In this sense, a Polaroid serves as a device that filters out one-half of the vibrations upon transmission of the light through the filter. When unpolarized light is transmitted through a Polaroid filter, it emerges with one-half the intensity and with vibrations in a single plane; it emerges as polarized light.

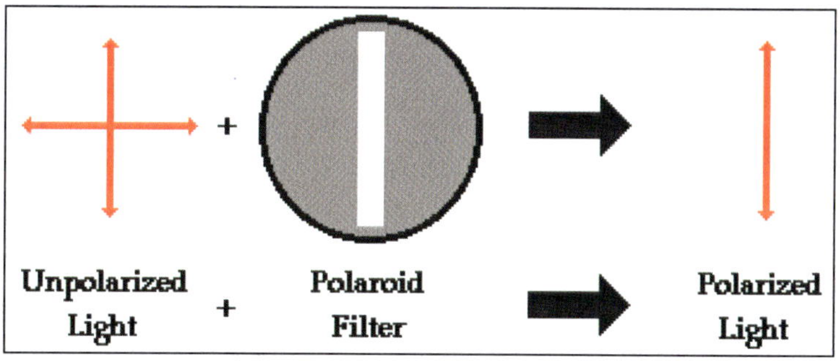

A Polaroid filter is able to polarize light because of the chemical composition of the filter material. The filter can be thought of as having long-chain molecules that are aligned within the filter in the same direction. During the fabrication of the filter, the long-chain molecules are stretched across the filter so that each molecule is (as much as possible) aligned in say the vertical direction. As unpolarized light strikes the filter, the portion of the waves vibrating in the vertical direction are absorbed by the filter. The general rule is that the electromagnetic vibrations that are in a direction parallel to the alignment of the molecules are absorbed.

The alignment of these molecules gives the filter a polarization axis. This polarization axis extends across the length of the filter and only allows vibrations of the electromagnetic wave that are parallel to the axis to pass through. Any vibrations that are perpendicular to the polarization axis are blocked by the filter. Thus, a Polaroid filter with its long-chain molecules aligned horizontally will have a polarization axis aligned vertically. Such a filter will block all horizontal vibrations and allow the vertical vibrations to be transmitted. On the other hand, a Polaroid filter with its long-chain molecules aligned vertically will have a polarization axis aligned horizontally; this filter will block all vertical vibrations and allow the horizontal vibrations to be transmitted.

Relationship Between Long-Chain Molecule Orientation and the Orientation of the Polarization Axis

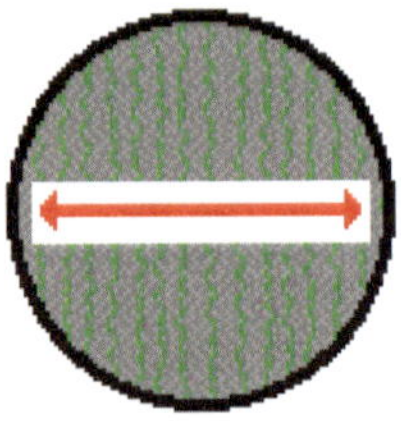

When molecules in the filter are aligned vertically, the polarization axis is horizontal.

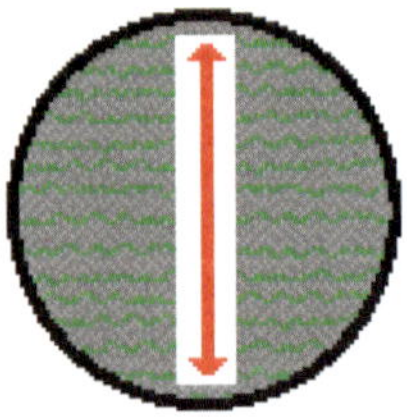

When molecules in the filter are aligned horizontally, the polarization axis is vertical.

Polarization of light by use of a Polaroid filter is often demonstrated in a Physics class through a variety of demonstrations. Filters are used to look through and view objects. The filter does not distort the shape or dimensions of the object; it merely serves to produce a dimmer image of the object since one-half of the light is blocked as it passed through the filter. A pair of filters is often placed back to back in order to view objects looking through two filters. By slowly rotating the second filter, an orientation can be found in which all the light from an object is blocked and the object can no longer be seen when viewed through two filters. What happened? In this demonstration, the light was polarized upon passage through the first filter; perhaps only vertical vibrations were able to pass through. These vertical vibrations were then blocked by the second filter since its polarization filter is aligned in a horizontal direction. While you are unable to see the axes on the filter, you will know when the axes are aligned perpendicular to each other because with this orientation, all light is blocked. So by use of two filters, one can completely block all of the light that is incident upon the set; this will only occur if the polarization axes are rotated such that they are perpendicular to each other.

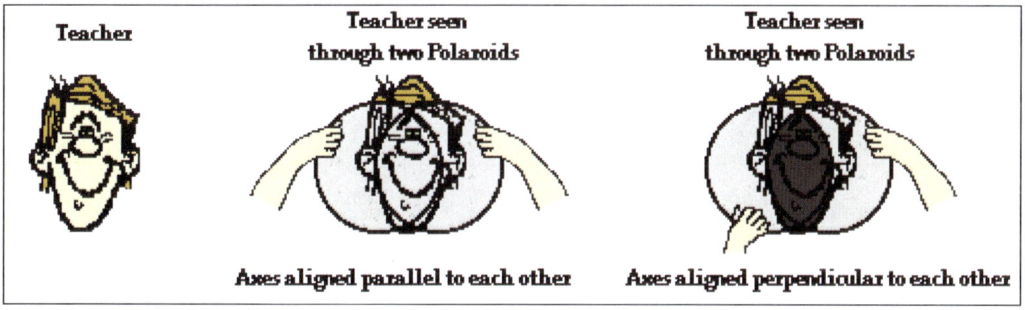

A picket-fence analogy is often used to explain how this dual-filter demonstration works. A picket fence can act as a polarizer by transforming an unpolarized wave in a rope into a wave that vibrates in a single plane. The spaces between the pickets of the fence will allow vibrations that are parallel to the spacings to pass through while blocking any vibrations that are perpendicular to the spacings. Obviously, a vertical vibration would not have the room to make it through a horizontal spacing. If two picket fences are oriented such that the pickets are both aligned vertically, then vertical vibrations will pass through both fences. On the other hand, if the pickets of the second fence are aligned horizontally, then the vertical vibrations that pass through the first fence will be blocked by the second fence. This is depicted in the diagram below.

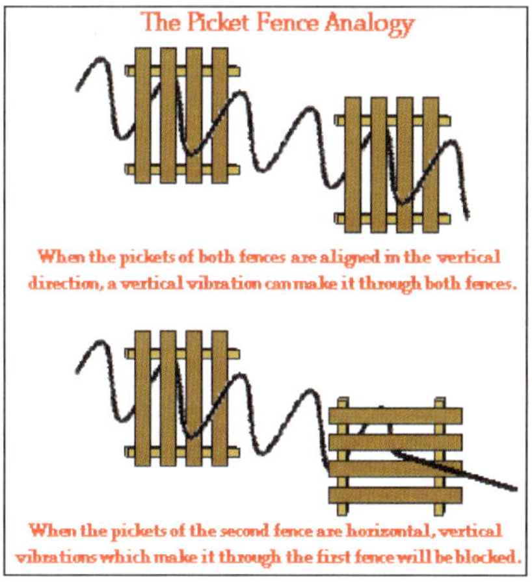

In the same manner, two Polaroid filters oriented with their polarization axes perpendicular to each other will block all the light. Now that's a pretty cool observation that could never be explained by a particle view of light.

Polarization by Reflection

Unpolarized light can also undergo polarization by reflection off of nonmetallic surfaces. The extent to which polarization occurs is dependent upon the angle at which the light approaches the surface and upon the material that the surface is made of. Metallic surfaces reflect light with a variety of vibrational directions; such reflected light is unpolarized. However, nonmetallic surfaces such as asphalt roadways, snowfields and water reflect light such that there is a large concentration of vibrations in a plane parallel to the reflecting surface. A person viewing objects by means of light reflected off of nonmetallic surfaces will often perceive a glare if the extent of polarization is large. Fishermen are familiar with this glare since it prevents them from seeing fish that lie below the water. Light reflected off a lake is partially polarized in a direction parallel to the water's surface. Fishermen know that the use of glare-reducing sunglasses with the proper polarization axis allows for the blocking of this partially polarized light. By blocking the plane-polarized light, the glare is reduced and the fisherman can more easily see fish located under the water.

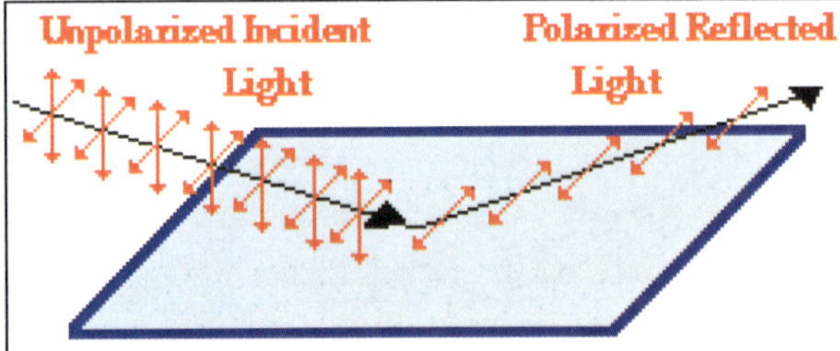

Reflection of light off of non-metallic surfaces results in some degree of polarization parallel to the surface.

Polarization by Refraction

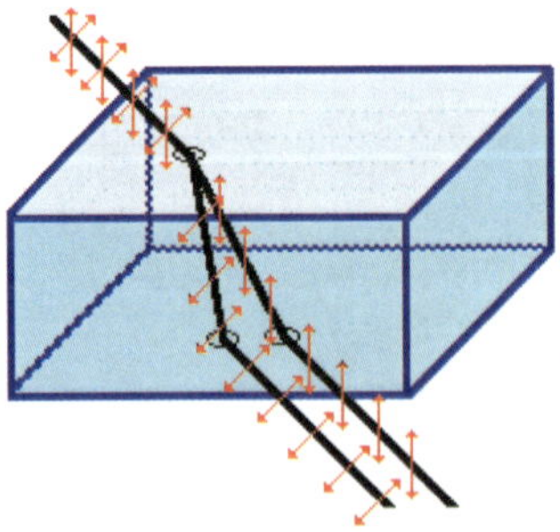

The two refracted rays passing through the Iceland Spar crystal are polarized with perpendicular orientations.

Polarization can also occur by the refraction of light. Refraction occurs when a beam of light passes from one material into another material. At the surface of the two materials, the path of the beam changes its direction. The refracted beam acquires some degree of polarization. Most often, the polarization occurs in a plane perpendicular to the surface. The polarization of refracted light is often demonstrated in a Physics class using a unique crystal that serves as a double-refracting crystal. Iceland Spar, a rather rare form of the mineral calcite, refracts incident light into two different paths. The light is *split* into two beams upon entering the crystal. Subsequently, if an object is viewed by looking through an Iceland Spar crystal, two images will be seen. The two images are the result of the double refraction of light. Both refracted light beams are polarized - one in a direction parallel to the surface and the other in a direction perpendicular to the surface. Since these two refracted rays are polarized with a perpendicular orientation, a polarizing filter can be used to completely block one of the images. If the polarization axis of the filter is aligned perpendicular to the plane of polarized light, the light is completely blocked by the filter; meanwhile the second image is as bright as can be. And if the filter is then turned 90-degrees in either direction, the second image reappears and the first image disappears. Now that's pretty neat observation that could never be observed if light did not exhibit any wavelike behavior.

Polarization by Scattering

Polarization also occurs when light is scattered while traveling through a medium. When light strikes the atoms of a material, it will often set the electrons of those atoms into vibration. The vibrating electrons then produce their own electromagnetic wave that is radiated outward in all directions. This newly generated wave strikes neighboring atoms, forcing their electrons into vibrations at the same original frequency. These vibrating electrons produce another electromagnetic wave that is once more radiated outward in all directions. This absorption and reemission of light waves causes the light to be scattered about the medium. (This process of scattering contributes to the blueness of our skies.) This scattered light is partially polarized. Polarization by scattering is observed as light passes through our atmosphere. The scattered light often produces a glare in the skies. Photographers know that this partial polarization of scattered light leads to photographs characterized by a washed-out sky. The problem can easily be corrected by the use of a Polaroid filter. As the filter is rotated, the partially polarized light is blocked and the glare is reduced. The photographic secret of capturing a vivid blue sky as the backdrop of a beautiful foreground lies in the physics of polarization and Polaroid filters.

Malus' Law

Suppose we have a second piece of Polaroid whose transmission axis makes an angle θ with that of the first one. The E vector of the light between the Polaroids can be resolved into two components, one parallel and one perpendicular to the transmission axis of the second Polaroid. If we call the direction of transmission of the second Polaroid y',

$$E_{x'} = E\sin\theta \text{ and } E_{y'} = E\cos\theta$$

Only the second, y' component is transmitted by the second Polaroid.

The intensity of light is proportional to the square of the electric field amplitude. Thus the intensity transmitted by both Polaroids can be expressed as:

$$I(\theta) \cong E_{y'}^2 = E^2 \cos^2 \theta$$

If $I_0 \cong E^2$ is the intensity between the two Polaroids, the intensity transmitted by both of them would be:

$$I(\theta) = I_0 \cos^2 \theta$$

This equation is known as Malus Law after its discoverer, E.L. Malus. It applies to any two polarizing elements whose transmission directions make an angle θ with each other. When two polarizing elements are placed in succession in a beam of light as described here, the first is called polarizer and the second is called analyzer. As you will see, no light reaches the photocell when the polarizer and analyzer are crossed (θ = 90°).

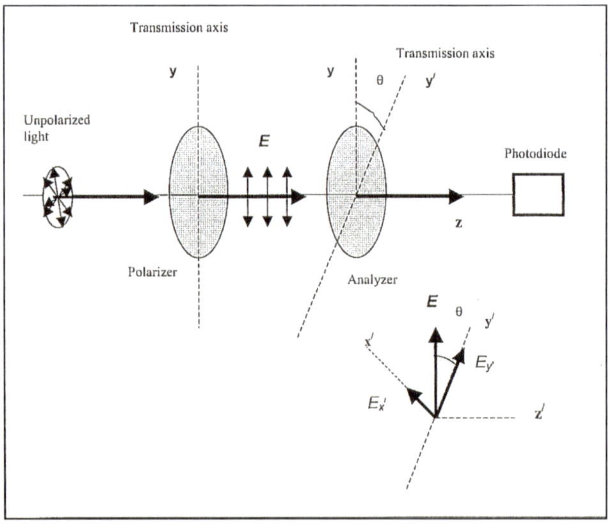

Figure: Two Polaroids whose transmission directions make an angle θ with each other.

Theory for Three Polarizers

Unpolarized light passes through 3 polarizers:

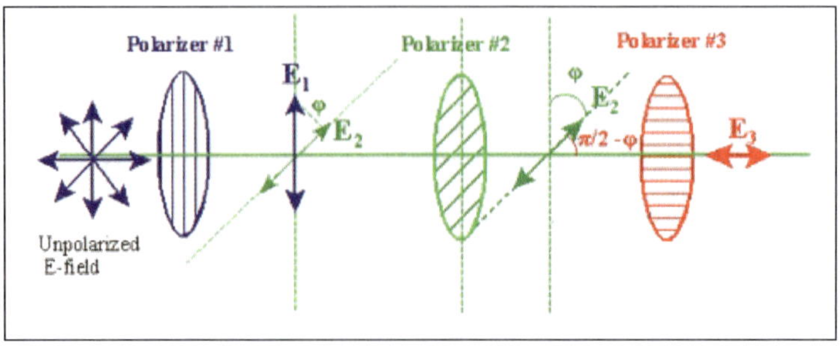

Figure: The electric field transmitted through three polarizers.

The first and last polarizers are oriented at 90° with respect to each other. The second polarizer has its polarization axis rotated an angle φ from the first polarizer. Therefore, the third polarizer is rotated an angle $\frac{\pi}{2}-\phi$ from the second polarizer. The intensity after passing through the first polarizer is I_1 and the intensity after passing through the second polarizer, I_2, is given by:

$$I_2 = I_1 \cos^2 \phi$$

The intensity after the third polarizer, I_3, is given by:

$$I_3 = I_2 \cos^2\left(\frac{\pi}{2}-\phi\right) = I_1 \cos^2 \phi \cos^2\left(\frac{\pi}{2}-\phi\right)$$

Rearranging Equation $I_3 = I_2 \cos^2\left(\frac{\pi}{2}-\phi\right) = I_1 \cos^2 \phi \cos^2\left(\frac{\pi}{2}-\phi\right)$, we obtain:

$$I_3 = \frac{I_1}{4}\sin^2(2\phi)$$

Because the data acquisition begins when the transmitted intensity through Polarizer 3 is a maximum, the angle Θ measured in the experiment is zero when the second polarizer is 45° from the angle φ. Thus the angle φ is related to the measured angle Θ by:

$$\phi = 45° + \Theta$$

Reflectance/Brewster's Angle

Consider unpolarized light incident on a surface separating air and glass or air and water. Define the plane of incidence, containing the incident, reflected and refracted rays as well as the normal to the surface.

When light is reflected from a flat surface, the reflected light is partially polarized. This is due to the fact that the reflectance of light R = (Reflected Intensity)/(Incident Intensity) depends on the polarization itself. The degree of polarization depends on the angle of incidence and the indices of refraction of the two media.

For reflection at an air-glass interface (indices of refraction n_1 for air and n_2 for glass), Fresnel equations give the reflection coefficients $r_{\parallel}, r_{\perp}$:

$$r_{\perp} = \frac{n_1 \cos\theta_1 - n_2 \cos\theta_2}{n_1 \cos\theta_1 + n_2 \cos\theta_2}$$

$$r_{\parallel} = \frac{n_1 \cos\theta_2 - n_2 \cos\theta_1}{n_1 \cos\theta_2 + n_2 \cos\theta_1}$$

where θ_1 is the angle of incidence, and θ_2 is the angle of refraction.

Above $r_{\parallel}$ and $r_{\perp}$ refer to the reflection coefficients for polarized light whose direction of polarization lie in the plane of incidence and perpendicular to the plane of incidence, respectively. Reflectance R (parallel or perpendicular) is defined as the square of the corresponding reflection coefficient: $R_{\parallel} = r_{\parallel}^2, R_{\perp} = r_{\perp}^2$

Note that θ_2 is not measured in this experiment and must be inferred from Snells law of refraction:

$$\frac{\sin\theta_2}{\sin\theta_1} = \frac{n_1}{n_2}$$

Figure shows initially unpolarized light incident at the polarizing angle θ_p, for which the reflected light is completely polarized with its electric field vector perpendicular to the plane of incidence. The electric field vector E of the incident wave can be resolved into two components: parallel to the plane of incidence and perpendicular to the plane of incidence.

Note that $R_{\parallel} = 0$ when $n_1 \cos\theta_1 = n_2 \cos\theta_2$, which leads to the definition of the Brewster angle (or polarizing angle):

$$\tan\theta_p = \frac{n_2}{n_1}$$

Here θ_p is the angle of incidence of unpolarized light which makes the reflected light completely polarized in the perpendicular direction to the plane of incidence.

When the angle of incidence of the initially unpolarized light is θ_p, the reflected and refracted rays are perpendicular to each other.

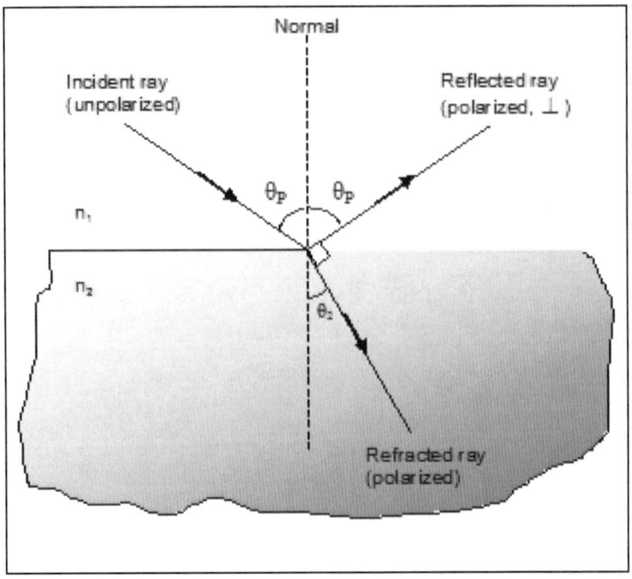

Figure: Unpolarized light incident at the polarizing angle.

Note: If the incident light has no component of E perpendicular to the plane of incidence, there is no reflected light.

Double Refraction

Double refraction, also called birefringence is an optical property in which a single ray of unpolarized light entering an anisotropic medium is split into two rays, each traveling in a different direction. One ray (called the extraordinary ray) is bent, or refracted, at an angle as it travels through the medium; the other ray (called the ordinary ray) passes through the medium unchanged.

Double refraction can be observed by comparing two materials, glass and calcite. If a pencil mark is drawn upon a sheet of paper and then covered with a piece of glass, only one image will be seen; but if the same paper is covered with a piece of calcite, and the crystal is oriented in a specific direction, then two marks will become visible.

The figure shows the phenomenon of double refraction through a calcite crystal. An incident ray is seen to split into the ordinary ray CO and the extraordinary ray CE upon entering the crystal face at C. If the incident ray enters the crystal along the direction of its optic axis, however, the light ray will not become divided.

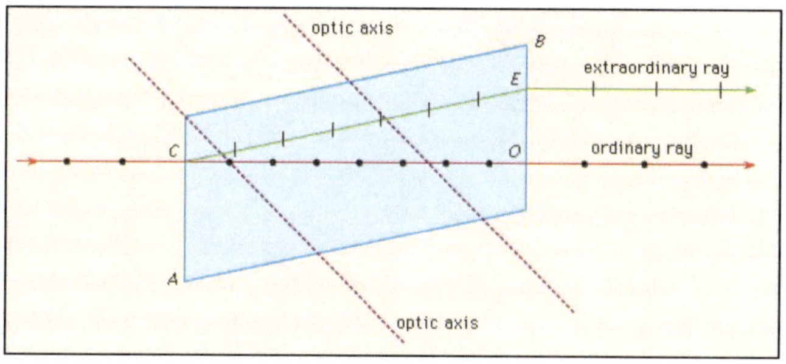

In double refraction, the ordinary ray and the extraordinary ray are polarized in planes vibrating at right angles to each other. Furthermore, the refractive index (a number that determines the angle of bending specific for each medium) of the ordinary ray is observed to be constant in all directions; the refractive index of the extraordinary ray varies according to the direction taken because it has components that are both parallel and perpendicular to the crystal's optic axis. Because the speed of light waves in a medium is equal to their speed in a vacuum divided by the index of refraction for that wavelength, an extraordinary ray can move either faster or slower than an ordinary ray.

All transparent crystals except those of the cubic system, which are normally optically isotropic, exhibit the phenomenon of double refraction: in addition to calcite, some well-known examples are ice, mica, quartz, sugar, and tourmaline. Other materials may become birefringent under special circumstances. For example, solutions containing long-chain molecules exhibit double refraction when they flow; this phenomenon is called streaming birefringence. Plastic materials built up from long-chain polymer molecules may also become doubly refractive when compressed or stretched; this process is known as photoelasticity. Some isotropic materials (e.g., glass) may even exhibit birefringence when placed in a magnetic or electric field or when subjected to external stress.

Uniaxial and Biaxial Optical Materials

Depending on the symmetry of the crystal structure, a crystalline optical material can be uniaxial or biaxial.

The simpler case is that of uniaxial crystals (examples: calcite, quartz, sapphire, $LiNbO_3$). Those have a so-called *optical axis*, and the refractive index for given wavelength depends on the relative orientation of electric field director and optical axis:

- If the electric field has the direction of the optical axis, one obtains the extraordinary index n_e. This is possible only if the propagation direction (more precisely, the direction of the k vector) is perpendicular to the optical axis. For the other polarization direction, one then obtains the ordinary index n_o.
- For propagation along the optical axis, the electric field can only be perpendicular to that axis, so that one obtains the ordinary index for any polarization direction. In that situation, no birefringence is experienced.
- For an arbitrary angle θ between propagation direction and optical axis, one can find two linear polarization directions exhibiting different refractive indices. The first one is perpendicular to the k vector and the optical axis; here, we have the ordinary index n_o, and such a wave is called an ordinary wave. The other polarization direction is perpendicular to that and to the k vector. The latter has a refractive index which is generally not the extraordinary index n_e, but a rather a mixture of n_e and n_o. This can be calculated with the following equation:

$$n = \frac{1}{\sqrt{\frac{\cos^2\theta}{n_o^2} + \frac{\sin^2\theta}{n_e^2}}}$$

The equation shows that for $\theta \to 0$ (i.e., propagation along the optical axis) we obtain $n \to n_o$, and the observed birefringence vanishes: one obtains n_o for any polarization direction.

One distinguishes positive and negative uniaxial crystals; in the former case, the extraordinary index is higher than the ordinary index.

For biaxial crystals (examples: mica, $CaTiO_3$ = perovskite, LiB_3O_5 = LBO, β-BaB_2O_4 = BBO), such calculations are substantially more complicated, at least for arbitrary propagation directions. There are three mutually orthogonal principal axes associated with different refractive indices. Frequently, however, one deals with cases where the propagation direction is in one of the planes spanned by the principal axes of index ellipsoid, and in such cases the calculation is again reasonably simple. This is usually the case in calculations for phase matching of nonlinear frequency conversion processes.

For optical fibers and other waveguides, the distinction between uniaxial and biaxial does not apply, since the propagation direction is essentially determined by the waveguide.

Consequences of a Polarization-dependent Refractive Index

The polarization dependence of the refractive index can have a variety of effects, some of which are highly important in nonlinear optics and laser technology:

- When a beam is refracted at the surface of a birefringent crystal, the refraction angle depends on the polarization direction. An unpolarized light beam can then be split into two linearly polarized beams when hitting surfaces of the material with non-normal incidence (*double refraction*). When some object, which is illuminated with unpolarized light, is viewed through a birefringent crystal (e.g. made of calcite), two images occur which are slightly displaced.
- If a linearly polarized laser beam propagates through a birefringent medium, there are generally two polarization components with different wavenumbers. Therefore, the polarization state changes during propagation due to the change in relative phase of the two linearly polarized components. This effect can be applied, for example, in birefringent tuners, because it is wavelength-dependent (even if the difference in refractive indices is not wavelength-dependent). It can also be power-dependent (→ *nonlinear polarization rotation*) through self- and cross-phase modulation, e.g. in an optical fiber, and this effect is sometimes used for passive mode locking of fiber lasers.
- Similarly, the polarization state of a laser beam in a laser crystal with thermally induced birefringence is distorted. The kind of distortion depends on the position, since the birefringent axis has a varying (e.g. always radial) orientation. This effect (combined with a polarizing optical element in the laser resonator) is the origin of depolarization loss.
- The birefringence of nonlinear crystal materials allows for birefringent phase matching of nonlinear interactions. Essentially, this means that birefringence compensates the wavelength dependence of the refractive index.
- For extraordinary waves, where the refractive index depends on the angular orientation, there is a spatial walk-off: the direction of power propagation is slightly tilted against that of the k vector.

Examples of Birefringence

In laser technology and nonlinear optics, the phenomenon of birefringence occurs mainly in the context of non-isotropic crystals:

- Some laser crystals (e.g. vanadate or tungstate crystals) are naturally birefringent. This is often helpful for obtaining a linearly polarized output without depolarization loss.
- All nonlinear crystals for nonlinear frequency conversion are birefringent.
- Birefringent crystals are also used for making polarizers.
- Although optical fibers are in most cases not birefringent by nature, birefringence is frequently encountered in fiber optics: some birefringence can result from bending (which also causes bend losses) and from random perturbations. Also, there are polarization-maintaining fibers.

Even in a naturally isotropic medium, birefringence can be induced e.g. by inhomogeneous mechanical stress. This can be observed e.g. by placing a piece of acrylic between two crossed polarizers: when stress is applied to the acrylic, one observes colored patterns resulting from the wavelength-dependent effect of stress-induced birefringence. Similar effects occur in bent optical fibers, and also due to thermal effects in laser crystals, which can lead to depolarization loss.

Straight optical fibers usually exhibit only a small degree of random birefringence, which can however scramble the polarization state of guided light over some propagation distance, e.g. 1 m. There are polarization-maintaining fibers, where a strong artificial birefringence can be used for suppressing such effects.

Quantifying Birefringence

The magnitude of birefringence can be specified in different ways:

- For an optical component with some birefringent, one can specify the retardance, which is the difference in phase shifts for the two polarization directions.
- For bulk optical materials, it is also common to consider the difference of refractive indices for the two polarization directions. The larger that difference, the larger the obtained retardance per millimeter of propagation length.
- For optical fibers and other waveguides, it is more appropriate to consider the difference of effective refractive indices. This is directly related to the difference in imaginary values of the propagation constants.
- Alternatively, one may specify the *polarization beat length*, which is 2π divided by the difference of the propagation constants. If waves with different polarization directions propagate together in the waveguide, their phase relation is restored after integer multiples of the propagation beat length.

Polarization Properties of Prisms and Reflectors

Prisms Producing Polarized Light

Certain types of crystals, such as calcite (Iceland spar or calcium carbonate) exhibit the property of double refraction or birefringence, as first observed in calcite by Erasmus Bartholinus in 1669. For the class of crystals called uniaxial, there is only one direction where all light rays travel along the same path at a constant velocity. This direction defines the optic axis or principal axis, and any plane that contains the optic axis is called a principal plane (sometimes called a principal section). The optic axis is not a specific line, but indicates a direction in the crystal where there is no double refraction. For all rays not traveling along the optic axis, the velocity is determined by a pair of refractive indices called the ordinary refractive index no and the extraordinary refractive index ne, and the path of an incident ray is split into two rays, the so-called o-rays and e-rays. Birefringence is specified by the number $\left(\mathrm{n_o} - \mathrm{n_e}\right)$. Moreover, these o-rays and e-rays are polarized and vibrate in mutually perpendicular planes. Only rays traveling parallel to the optic axis will not be split, and no is therefore assigned to this direction. One way to represent this refractive index variation is by use of the indicatrix. Figure (a) shows a positive uniaxial indicatrix in the shape of an oblate spheroid, where $n_e > n_o$, and Figure (b) shows a negative uniaxial indicatrix in the shape of a prolate spheroid, where $n_o > n_e$. Both have circular symmetry in planes normal to the optic axis, and when the indicatrix has a spherical shape, $n_e = n_o$, and the crystal is isotropic.

Nicol Polarizing Prism

One of the first prism polarizers to utilize a birefringent crystal was developed by William Nicol in 1828 and is known as the Nicol prism. The Nicol prism shown in figure is constructed from negative uniaxial calcite, where $\mathrm{n_o} = 1.6584$ and $\mathrm{n_e}$=1.4864 for $\lambda = 589.3$ nm. Calcite is a widely used material because of its clarity, stability, high spectral transmission range (200–5000 nm), and high birefringence. Two triangular sections are optically coupled at the hypotenuse by a thin coating of optically clear cement such as Canadian balsam $\left(\mathrm{n_{cement}} \approx 1.54\right)$, with the optic axis direction as shown. An incident unpolarized ray is split at the entrance surface, with both rays becoming linearly polarized. By controlling the incident angle of the rays at the interface, the o-ray can undergo total internal reflection (TIR), where $\mathrm{I^\circ_{crit}} = \arcsin\left(\mathrm{n_{cement}} / \mathrm{no}\right) \approx 68$ deg. Since $\mathrm{n_{cement}} > \mathrm{n_e}$, the e-ray is always transmitted and exits the prism as linearly polarized light. This separation of o-rays and e-rays by TIR is a useful technique that is used in other types of polarizing prisms. Although the exit ray is parallel to the incident ray, there is a slight lateral displacement (noncollinear), the angular field is limited, and the interface cement will suffer damage at high power levels.

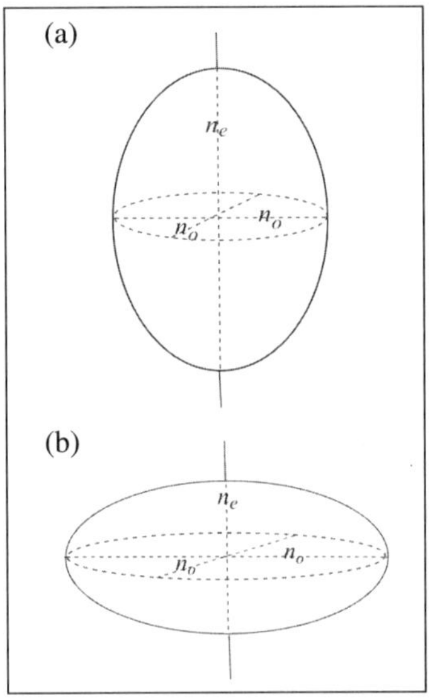

(a) Positive uniaxial indicatrix ($n_e > n_o$). (b) Negative uniaxial indicatrix ($n_o > n_e$).

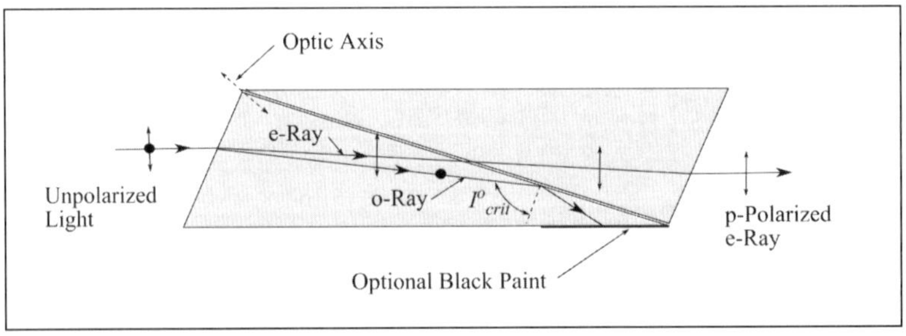

Nicol prism polarizer made of calcite, $n_o = 1.6584$, $n_e = 1.4864$.

Glan–Foucault Polarizing Prism

One of several Glan-type polarizing prisms is the Glan–Foucault prism, shown in the figure. The calcite prisms are air spaced at the interface, and each optic axis is perpendicular to the plane of reflection. There is no separation of the ray paths in the first prism section, but the o-ray moves slower in the first section and undergoes retardance with respect to the e-ray. Again, TIR is used to separate the o-ray from the e-ray, and s-polarized light is emitted from the exit face. The o-ray is usually absorbed by blackening the side face. The field of view is determined by TIR failure of the o-ray at the glass–air interface 2, or TIR of the e-ray at this interface. For calcite, $n_o = 1.6557$ and $n_e = 1.4852$ at $\lambda = 630$ nm. The corresponding critical angles are $I^o_{crit} = 37.16$ deg and $I^e_{crit} = 42.32$ deg. As shown in figure, the maximum angle of incidence I_1^0 for the o-ray at entrance surface 1 is estimated by $I^0_{1\ max} = \arcsin[n_o \sin(38.5\text{ deg} - I^o_{crit})] \approx 2.4$ deg. The maximum angle of incidence I_1^e for the e-ray is estimated by I1 e max = $\arcsin[n_o \sin(I^o_{crit} - 38.5\text{ deg})] \approx 6.3$ deg. This results in a narrow asymmetric field of view about the central axis in the tangential plane. A nominal angular field is given as 6 deg at $\lambda = 633$ nm by a commercial supplier of the Glan–Foucault prism, United Crystals Company.

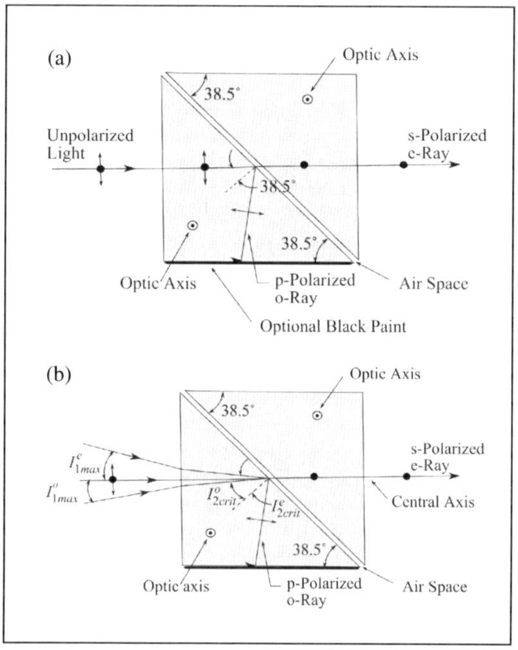

(a) Glan–Foucault prism polarizer made of calcite, I_{crit} (no) = 37.1 deg, $I_{crit}(n_e)$ = 42.3 deg. (b) Asymmetric field of view of Glan–Foucault prism polarizer.

Other specifications are a damage threshold of 30 W/cm^2 continuous wave (CW) or 300 W/cm^2 pulsed laser radiation and transmittance of s-polarized light > 60% at λ = 633 nm.

Glan–Thompson Polarizing Prism

The Glan–Thompson prism shown in the figure uses two cemented calcite prisms with each optic axis perpendicular to the plane of reflection. Using TIR separation at the glass–optical cement interface, s-polarized light is transmitted, while the reflected p-polarized light is absorbed by a blackened side face. The transmission of s-polarized light is > 90%, and the angular field is approximately doubled to about 12 deg compared to the Glan–Foucault prism, but the Glan–Thompson prism can accept only up to 8 W/cm^2 CW or 100 W/cm^2 pulsed radiation due to the lower damage threshold of the optical cement.

Glan–Taylor Polarizing Prism

The last of the Glan group to be described here is the Glan–Taylor prism, shown in the figure. Two air-spaced calcite prisms are oriented with both optic axes parallel to the plane of reflection and parallel to the entrance and exit faces. Using TIR separation, p-polarized light is transmitted, while the reflected s-polarized light is either absorbed by a blackened side face or emitted through a clear exit window. The transmission of p-polarized light is > 85%, and the angular field is about 6 deg. It can accommodate the highest radiation level of the Glan group—up to 30 W/cm^2 CW or 500 W/cm^2 pulsed radiation. A modified form of the Glan–Taylor prism, shown in figure, can produce orthogonal s-polarized and p-polarized output beams. In addition, if the angle of incidence at the interface is close to Brewster's angle, there will be little reflection of p-polarized light. However, the intensity of the s-polarized reflected beam will be much less than the transmitted p-polarized beam.

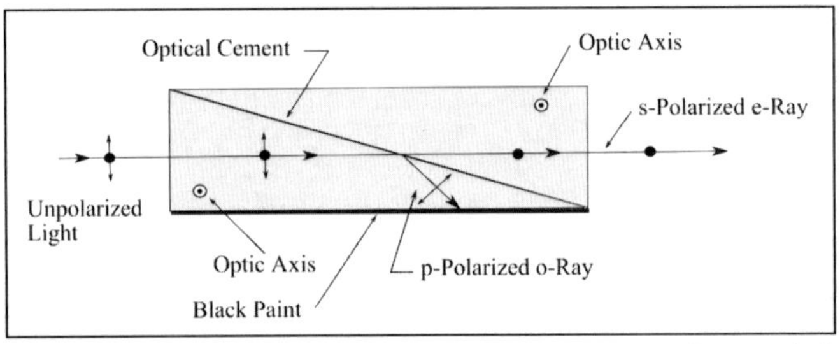

Glan–Thompson prism polarizer made of calcite, $I_{crit}(n_o) = 37.1$ deg, $I_{crit}(n_e) = 42.3$ deg.

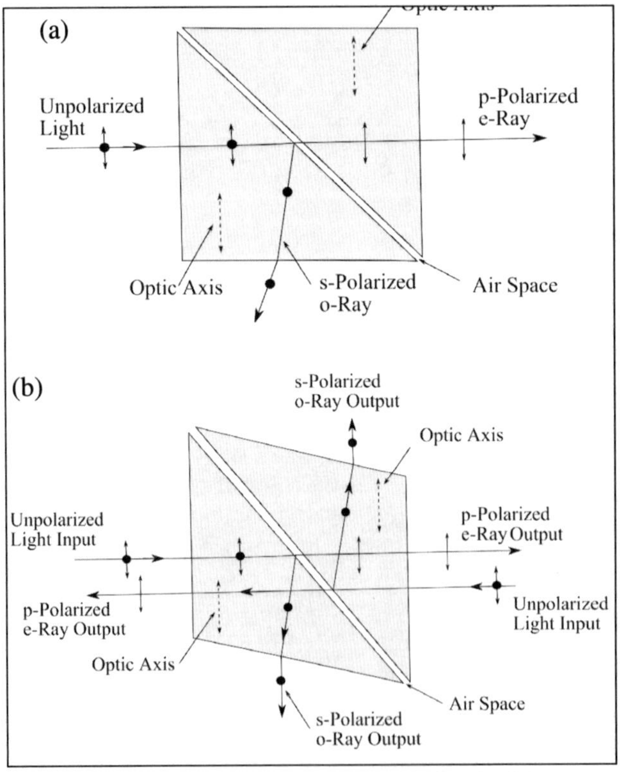

(a) Glan–Taylor prism polarizer made of calcite.
(b) Glan–Taylor prism polarizer having orthogonal outputs made of calcite.

Beam-displacing Polarizing Prism

Figure shows a cleaved calcite rhomb where the optic axis is inclined in the principal section. Each side has corner angles $\alpha = 78.08$ deg and $\beta = 101.92$ deg. The optic axis direction is determined by equally trisecting a β-β-β oblique corner of the crystal. A ray enters at the edge of the principal section. The undeviated s-polarized o-ray vibrates perpendicular to the principal section, and the deviated p-polarized e-ray vibrates in the principal section, where both rays lie in the principal section. The exiting p-polarized ray is displaced and parallel to the exiting s-polarized ray. Typical exit ray separation is nominally 4 mm. This polarizing prism has the advantage that both exit beams are completely polarized and of equal intensity, although obviously the entrance beam diameter must be small. As in figure, a calcite prism pair can produce a variable beam displacement between the p-polarized and s-polarized rays. Here, one wedge prism is slid relative to another wedge prism.

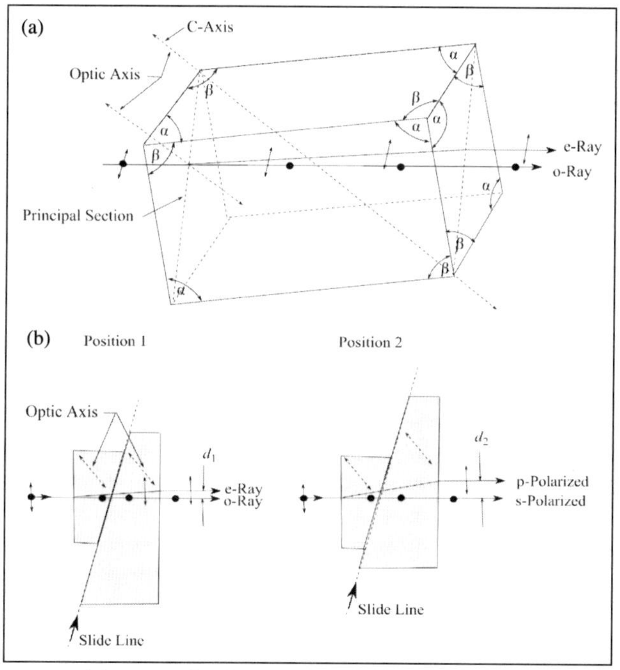

(a) Beam-displacement prism polarizer. (b) Variable beam-displacement prism polarizer.

Wollaston Polarizing Prism

Another type of beam-splitting polarizing prism is the Wollaston prism, (for William Hyde Wollaston), shown in the figure. It usually consists of two calcite rightangle prisms optically cemented together at the hypotenuse. The optic axis of each section is orthogonal to that of the other section. An unpolarized ray traversing the first prism section is not split, but the o-ray is retarded with respect to the e-ray. The o-ray vibrates parallel to the optic axis and the e-ray perpendicular to the optic axis. Upon entering the second section, the o-ray becomes the e-ray, and vice versa. The o-ray, now slower, is bent toward the interface normal, and the e-ray is bent away from the interface normal. Prisms with a deviation angle δ from about 5 to 45 deg between the exit beams can be obtained, depending on the right-angle prism base angles. For very high-power applications, the prisms may not be cemented together, resulting in a reduction of transmission.

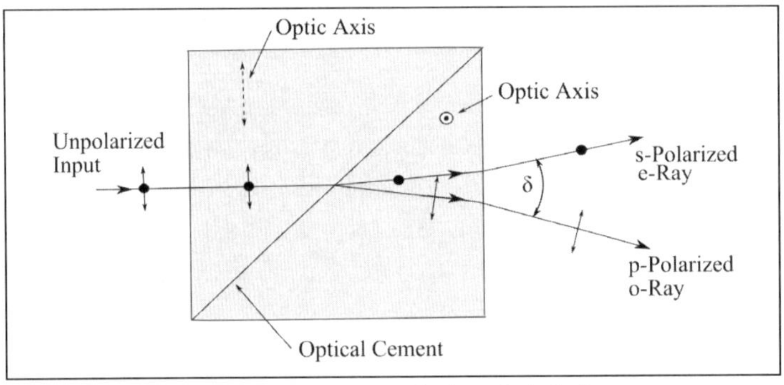

Wollaston prism polarizer (calcite).

Nomarski Polarizing Prism

The Nomarski prism, named for Georges Nomarski, is a modified Wollaston prism. The optic axis of the first right-angle calcite prism is skewed as shown, while the optic axis of the second prism is oriented the same as for the Wollaston prism. This angled optic axis causes the ordinary and extraordinary rays to intersect outside the prism, forming an interference plane. The resulting phase shifts can be detected by an analyzer. The exact distance of this interference plane from the prism is determined by the angle of the skewed optic axis and is set by the manufacturer. Normarski prisms are used in differential interference contrast (DIC) microscopes.

Rochon Polarizing Prism

Related to the Wollaston polarizing prism, the Rochon prism (for Alexis Marie Rochon) has the optic axis of the first calcite prism section in the direction of the incident ray, and there is no distinction between the o-ray and the e-ray in this section. The split at the second section interface produces no deviation of the s-polarized o-ray, while the p-polarized e-ray is deviated from the interface normal. The first calcite section can be replaced by a more robust isotropic glass section, choosing a glass with a refractive index and dispersion close to either of the refractive index values of calcite. Glass-calcite Rochon prisms are commercially available using FK5 glass ($n_d = 1.4875$, $\nu_d = 70.41$), which is close to n_e and the dispersion of calcite. Typical beam-deviation angles are 5, 10, and 15 deg.

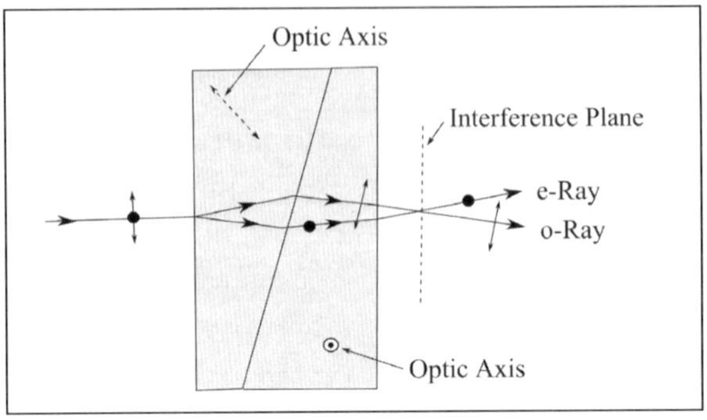

Nomarski prism polarizer (calcite).

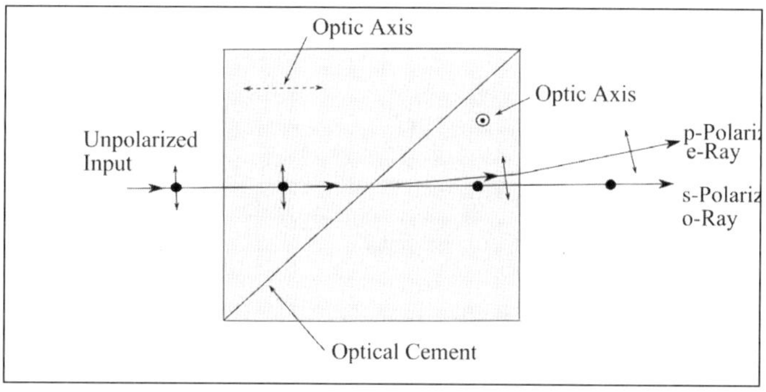

Rochon prism polarizer (calcite).

Other double-refracting crystals used in commercially available polarizing prisms are crystal quartz (circular and low birefringence), alpha-BBO (n_e = 1.6021, n_o = 1.6776 at λ = 552 nm), YVO4 (n_e = 2.2154, n_o = 1.9929 at λ = 630 nm), magnesium fluoride, and titanium dioxide. Polarizing prisms' surfaces are usually antireflection coated.

MacNeille Polarizing Beamsplitter Cube

Another method to produce polarized light is by the deposition of alternating highand low-index film layers, as described by MacNeille. Figure illustrates a seven-layer transparent thin-film stack having refractive indices n1 and n_2, deposited between transparent bulk material having a refractive index n3, where $n_1 \gg n_2$, and $n_1 > n_3$. All internal rays within the layers hit the next layer at Brewster's angle. A fraction of the reflected light at each layer interface is therefore completely s-polarized. If we choose n_1 = 2.3 (zinc sulfide) and n_2 = 1.38.

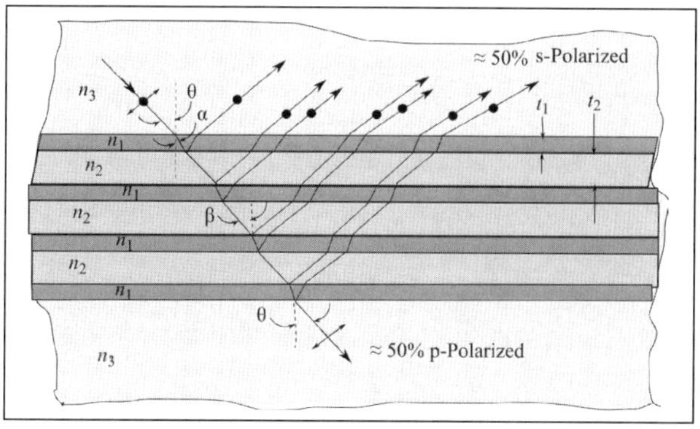

Polarizing thin-film stack.

(magnesium fluoride), then Brewster's angles α and β are given by,

$$\alpha = \arctan\left(\frac{n_2}{n_1}\right) = 31.0\,\text{deg},$$

$$\beta = \arctan\left(\frac{n_1}{n_2}\right) = 59.0\,\text{deg},$$

where, $\alpha + \beta = 90$ deg and Snell's law is satisfied at each layer interface.

A useful incident angle from the bulk material to the first layer is $\theta = 45$ deg. The required refractive index of the bulk material is then calculated from Snell's law:

$$n_3 = n_1 \left[\frac{\sin\alpha}{\sin(45\,\text{deg})} \right] \approx 1.67$$

Since θ between n_3 and n_2 (53.9 deg) is not Brewster's angle, this incident ray is not completely s-polarized on reflection.

To maximize the intensity of the reflected s-polarized ray at each layer, the layer thickness is controlled such that the ray reflected from the next layer is in phase with the incident ray. To achieve this, the physical thicknesses t_1 and t_2 of the layers are controlled to be,

$$t_1 = \frac{\lambda}{4\sqrt{\left(n_1^2 + n_2^2\right)/n_1^2}},$$

$$t_2 = \frac{\lambda}{4\sqrt{\left(n_1^2 + n_2^2\right)/n_2^2}},$$

Where λ is the wavelength of the incident light, nominally 550 nm. For these seven layers, approximately 50% of the incident light is reflected as s-polarized, while the other half is transmitted as p-polarized light.

Figure shows a 50R/50T polarizing beamsplitter (PBS) cube, where the deposited layers lie on the hypotenuse of a right-angle prism, and another rightangle prism is coupled to the hypotenuse using a thin coating of optical cement having a refractive index close to n_3. From Eq. $n_3 = n_1 \left[\frac{\sin\alpha}{\sin(45\,\text{deg})} \right] \approx\sim 1.67$, a suitable material for the cube would be SF5 glass ($n_d = 1.673$). Both the reflected s-polarized light and the p-polarized transmitted light are at least 95% polarized over the visible spectrum, and the beamsplitter is usable for 40 deg $\leq \theta \leq$ 50 deg, or $\pm$ 5 deg from the ideal incident angle at the interface. The extinction ratio is the ratio of the transmitted or reflected primary polarization component to the opposite polarization component. It is possible to increase the angular field of MacNeille PBS cubes by modifying the beam-splitting coating, albeit with a reduced usable wavelength range. Modern commercial PBS cubes of a modified MacNeille design can achieve an input f/# down to $\approx$ f/2.5 and a transmission extinction ratio $\approx$ 1,000:1. The minimum working f/# of a PBS cube for projection display applications, without noticeable loss of contrast, has been stated to be $\approx$ f/3.3.

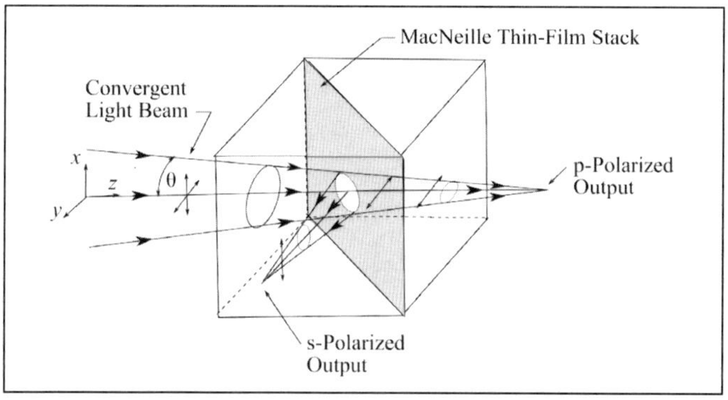

Figure: MacNeille 50R/50T PBS cube in convergent beam of half-angle θ.

Birefringent Multilayer Reflective Polarizing Film

A type of reflecting polarizer film has been developed by 3M that uses a multilayer stack of biaxial birefringent polymer layers. It is designed to produce high reflectance for light with its plane of polarization parallel to one axis, and high transmission for light with its plane of polarization parallel to a second axis, both over a wide range of incident angles. A biaxial birefringent material, where the refractive indices differ along all three axes, can be produced by stretching the multilayer stack in one direction (uniaxial stretching). Figure shows a single interface between layers for a biaxial birefringent film, and the associated refractive indices. For light incident in the x-y stretch plane, $n_{10} = n_{1x}$, $n_{20} = n_{2x}$ for p-polarized light, and $n_{10} = n_{1y}$, $n_{20} = n_{2y}$ for s-polarized light. For light incident in the y-z nonstretched plane, $n_{10} = n_{1y}$, $n_{20} = n_{2y}$ for p-polarized light, and $n_{10} = n_{1x}$, $n_{20} = n_{2x}$ for s-polarized light. The x direction is the extinction direction, and the y direction is the transmission direction. Typical values are $n_{1x} = 1.88$, $n_{1y} = 1.64$, n_{1z} = variable, $n_{2x} = 1.65$, n_{2y} = variable, and n_{2z} = variable. For the large index differential of 1.88–1.65 = 0.23 in the stretch direction, there is a high reflectance of s-polarized light for a stack of hundreds of layers, and the angular transmission of p-polarized light depends on the n_{1z}/n_{2z} index ratio.

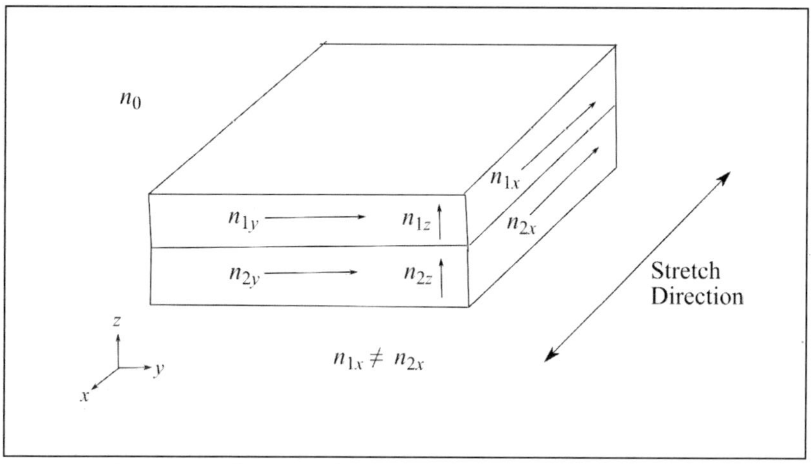

Figure: Two-layer single interface biaxial birefringent film.

Permissions

All chapters in this book are published with permission under the Creative Commons Attribution Share Alike License or equivalent. Every chapter published in this book has been scrutinized by our experts. Their significance has been extensively debated. The topics covered herein carry significant information for a comprehensive understanding. They may even be implemented as practical applications or may be referred to as a beginning point for further studies.

We would like to thank the editorial team for lending their expertise to make the book truly unique. They have played a crucial role in the development of this book. Without their invaluable contributions this book wouldn't have been possible. They have made vital efforts to compile up to date information on the varied aspects of this subject to make this book a valuable addition to the collection of many professionals and students.

This book was conceptualized with the vision of imparting up-to-date and integrated information in this field. To ensure the same, a matchless editorial board was set up. Every individual on the board went through rigorous rounds of assessment to prove their worth. After which they invested a large part of their time researching and compiling the most relevant data for our readers.

The editorial board has been involved in producing this book since its inception. They have spent rigorous hours researching and exploring the diverse topics which have resulted in the successful publishing of this book. They have passed on their knowledge of decades through this book. To expedite this challenging task, the publisher supported the team at every step. A small team of assistant editors was also appointed to further simplify the editing procedure and attain best results for the readers.

Apart from the editorial board, the designing team has also invested a significant amount of their time in understanding the subject and creating the most relevant covers. They scrutinized every image to scout for the most suitable representation of the subject and create an appropriate cover for the book.

The publishing team has been an ardent support to the editorial, designing and production team. Their endless efforts to recruit the best for this project, has resulted in the accomplishment of this book. They are a veteran in the field of academics and their pool of knowledge is as vast as their experience in printing. Their expertise and guidance has proved useful at every step. Their uncompromising quality standards have made this book an exceptional effort. Their encouragement from time to time has been an inspiration for everyone.

The publisher and the editorial board hope that this book will prove to be a valuable piece of knowledge for students, practitioners and scholars across the globe.

Index

O

P

Q

R

S

T

V

W